S0-BHZ-480

NFPA
INSPECTION MANUAL

NFPA
INSPECTION MANUAL

Fifth Edition

A guide to property inspection for fire protection and life safety.

Charles A. Tuck, Jr., Editor

National Fire Protection Association
Quincy, Massachusetts

Fifth Edition

First Printing October 1982

Copyright © 1950, 1959, 1962, 1970, 1976, 1982
National Fire Protection Association
All rights reserved

*National Electrical Code, National Fire Codes, Life Safety
Code*, and *101* are registered trademarks of the National Fire
Protection Association.

NFPA No. SPP-11C
ISBN 0-87765-239-2
Library of Congress No. 76-5194
Printed in U.S.A.

TABLE OF CONTENTS

INTRODUCTION

The fundamental reason for conducting fire inspections is to limit the risk of life and property losses from fire by identifying and causing the correction of those conditions which contribute to the occurrence and spread of fire. More specifically, the approach taken to inspection depends upon the reasons for making the inspection and the responsibilities of the inspecting authority.

Code enforcement inspections are conducted to ensure that the premises, facilities, equipment, processes, operations, and fire protection of a property comply with code requirements that have been adopted as a matter of law. They include plans review and inspections during construction as well as periodic inspections to monitor conditions. Such reviews and inspections might be conducted by the fire marshal's office or by the fire prevention bureau.

Prefire planning inspections are conducted by local fire fighting units and their district supervisors (chiefs) to acquaint themselves with building features and facilities, exposures, processes, hazardous materials, and fire fighting provisions that are characteristic of the property. The primary purpose of such inspections is to aid fire fighting personnel in developing a plan of attack in the event a fire does occur. However, inspection personnel may also order the immediate correction of extreme hazards and report all hazards and violations to the fire marshal's office or the bureau of fire prevention.

Private inspections are conducted to minimize the likelihood of a fire occurring or, if a fire should occur, ensuring that its duration and spread will be limited. Such inspections may involve the application of standards and practices that do not have the weight of law as well as those that do. Emphasis is on the preservation of life and property through the practice of prudent, though not necessarily legally required, fire protection measures. This type of inspection may be conducted by in-house personnel, insurance inspectors, or private inspection and engineering firms. Private inspections made by in-house personnel generally occur more frequently than those made by an outside authority, agency, or contractor. In addition, private inspections play an important role in ensuring compliance with federal requirements, such as the Occupational Safety and Health Act (OSHA), particularly as increased emphasis is now being placed on voluntary self-supervision.

This manual deals primarily with private inspections, but will also touch upon code enforcement and prefire planning inspections to the extent that their requirements overlap those of private inspections.

Private engineering and inspection organizations may also survey and map a property during the initial inspection and after alterations or additions and occupancy changes have occurred. Surveying and mapping are beyond the scope of this manual.

BIBLIOGRAPHY

McKinnon, G. P. ed., *Fire Protection Handbook*, 15th ed., National Fire Protection Association, Quincy, MA, 1981. The elements of surveying and mapping are discussed in Section 20, Chapter 1.

Chapter 1

THE FIRE INSPECTOR

Not just anyone can be a successful fire inspector. Individuals selected for inspection duties must enjoy good health, possess good communications skills, and be knowledgeable about the property and its facilities, contents, operations, and fire protection provisions. They need to have good judgment and an understanding of what they are to accomplish.

PHYSICAL CONDITION

Inspection routines may require inspectors to climb ladders, crawl into confined spaces, lift heavy objects (manhole covers, for instance), turn valve wheels, or push heavy doors. Therefore, in order to gain access to, and inspect, all parts of the premises, they should have the strength, agility, and stamina to do these things.

COMMUNICATIONS SKILLS

A vital part of the inspection process is the relating of problems and their solutions to the appropriate parties and the recording of conditions found and actions taken. The inspector will be called upon to communicate with management and peers, contractors and vendors, and representatives of the fire service and the insurance industry. Therefore, the inspector must be able to communicate clearly with others, both orally and in writing.

KNOWLEDGE

The breadth of knowledge required of fire inspectors is determined by the physical characteristics of the facilities to be inspected, the materials contained in them, and the operations they house.

Construction

With the introduction of new processes or product lines, building configurations are subject to alteration. During periods of construction, renovation, or demolition, properties are especially vulnerable to fire. The inspector should have sufficient knowledge of building construction and materials to be able to recognize and remedy potentially hazardous conditions as well as to be able to recommend temporary protective steps that can be taken.

Building Services

Not to be overlooked are building services, which may introduce fire hazards. In this regard, inspectors should know about the hazards of such building services as electrical systems, heating systems, air conditioning and ventilating systems, waste handling systems, and materials handling systems.

Hazardous Materials

The fire inspector should be familiar with the characteristics, hazards, proper handling and storage, and protection of a wide variety of hazardous materials that may be encountered on the premises. Typical hazardous materials include flammable and combustible liquids; compressed or liquefied flammable gases; explosives; corrosives; reactive materials; unstable materials; toxic materials; oxidizers; radioactive materials; natural and synthetic fibers; combustible metals; and combustible dusts.

Process Hazards

Some industrial processes introduce unusual hazards. The fire inspector should know what they are, how to minimize them, and what form of fire protection is suitable under a given set of circumstances.

Fire Protection Equipment

A variety of fire protection equipment may be provided on the premises. Most common are portable fire extinguishers, sprinkler systems, and standpipe and hose systems. Some areas and processes may be protected by special fixed extinguishing systems. For example, a halogenated extinguishing agent system may be used in a computer room, or a dipping or coating process may be protected by a system using carbon dioxide, dry chemical, or foam as the extinguishing agent. The inspector should have an understanding of the application and operation of any extinguishing equipment provided on the premises.

A property may also be equipped with heat, smoke, and flame detection equipment to provide early warning of developing fire. Fire alarm systems and devices may be provided to alert the occupants and summon the fire department. The fire inspector should be acquainted with the purpose and operation of such devices and systems.

BIBLIOGRAPHY

NFPA 1031, *Standard for Professional Qualifications for Fire Inspector, Fire Investigator and Fire Prevention Education Officer.* This standard prescribes the skills and knowledge required for the qualifications of three levels of fire service fire inspector. It can be used as the basis for developing qualifications for private fire inspectors.

Chapter 2

INSPECTION PROCEDURE

Inspectors are part detective, part reporter, part technical consultant, part missionary, and part salesperson. An inspection should produce a property that is safer because the inspection was made, inspire an improved attitude toward fire prevention by management and employees, and provide a record of the findings and actions resulting from the inspection.

INSPECTORS' EQUIPMENT

The inspector should be provided with some visible means of identification, such as an identification card or badge. In order to conduct the fire inspection safely and efficiently, the inspector should be properly equipped.

Personal Equipment

When dirty or hard-to-reach areas are inspected, coveralls and perhaps overshoes may be needed to protect the inspector's street clothes. Boots may be necessary when conducting waterflow tests.

In some environments — heavy industry or construction, for example — the use of safety equipment such as a hard hat, safety shoes, safety glasses, and gloves may be in order. Ear protection may be required when the inspector is exposed to a noisy environment. In still other instances, it may be necessary for the inspector to use respiratory protection devices such as a dust mask or self-contained breathing apparatus.

Aids to Inspection

The basic tools of a fire inspector are a working flashlight, a notebook or clipboard in which to make sketches or record observations not provided for elsewhere, report forms, and a pen or pencil. If a sketch is to be drawn, the accuracy of dimensions measured by pacing is often adequate. When greater accuracy is required, the inspector will find a 6-foot rule or a 50-foot measuring tape to be helpful.

More sophisticated equipment that may be needed on an occasional basis would include gages and connections for making waterflow measurements or a combustible gas detector for testing a potentially hazardous environment.

PREPARATION

Resident inspectors of a property who inspect the premises continuously probably will need little in the way of preparation after they have made a few inspections. They may need only to remind themselves of chronic trouble areas to be watched carefully.

Nonresident inspectors, however, should prepare themselves by reviewing previous inspection reports and surveys, learning about the operations and activities carried out on the premises, and preparing a list of the more important points to be investigated before starting the inspection.

Small-unit properties, such as one-story mercantile establishments or gasoline filling stations, may require little preparatory work by the inspector. On the other hand, larger more complex properties should be reviewed, at least in the inspector's mind, before the inspection is started.

THE APPROACH

Inspections are usually made during normal business hours, and advance arrangements may be made for inspections at other hours. For example, the inspector may visit the

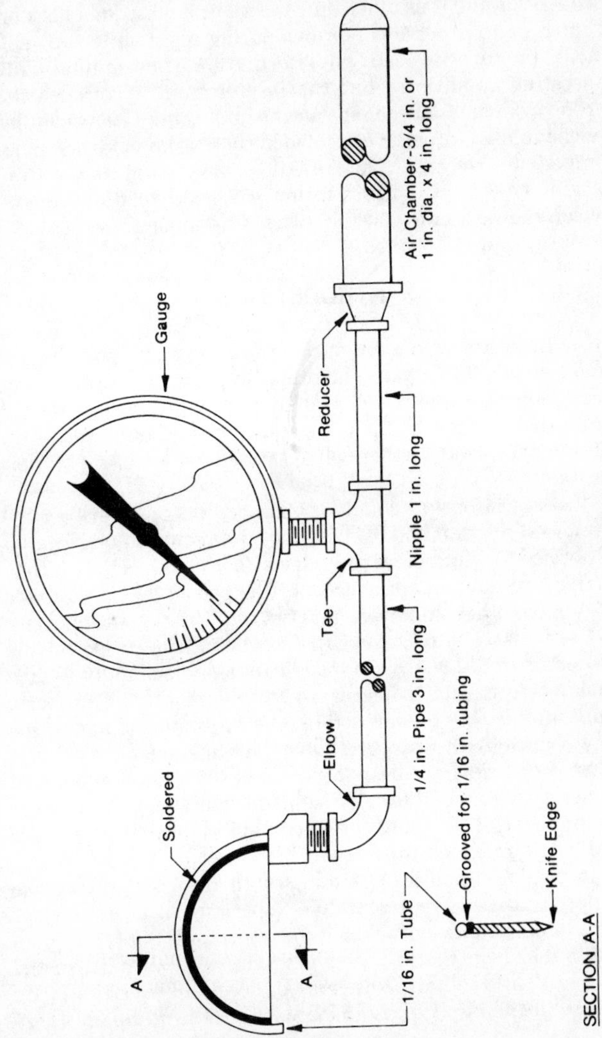

Figure 2-1. Pitot tube assembly.

property at night or other time the plant is closed to check on guards or to observe conditions during night shifts. An element of surprise can be effective in determining true operating conditions, but the inspector must consider the property owner's or manager's schedule. It may be prudent to reschedule an inspection, provided there is no evidence of an immediate fire or life hazard, if an inspection at the time originally chosen is likely to harm a good relationship between the inspector and the owner or manager.

INTRODUCTION

An inspector who is not part of the occupancy's staff should make an effort to create a favorable impression in order to ensure cooperation and courteous treatment. Inspectors should enter the premises by the main entrance, seek out someone in authority, introduce themselves, and state the nature of their business.

Visiting inspectors should ask for permission to inspect the premises, not demand it. They have no reason to be irritated if obliged to wait before receiving attention, especially if they arrived without an appointment. The person they need to see may have other important matters requiring attention first.

Particularly during the first inspection of a property, the inspector would be wise to spend a reasonable amount of time making sure that whoever is in charge of the property understands why the inspection is being made and answering any questions the property owner or manager may have. Most properties have been inspected at some time. Records of such inspections can usually be made available by the owner or manager. The records often contain plans that could save the inspector much time or work.

Inspections should be conducted in the company of the property owner's representative, especially in a property or class of occupancy with which the inspector is not familiar. A guide will help the inspector gain access to all parts of the property and obtain answers to necessary questions. Of course, if the inspector is an employee of the occupant, there will be no need for a guide once the initial inspection has been made.

THE INSPECTION SEQUENCE

In a large plant, the inspector may wish to start by touring the outside of the plant to observe the relationships of buildings to one another and to adjacent properties. If a ground plan of the property is available, it will be helpful to the inspector in visualizing the layout of the premises. It may also be helpful to obtain an overall view of the property from the top of the tallest building.

Whether a building is inspected from top to bottom or from bottom to top is of little consequence; it is the inspector's choice. What is important is that the inspection be done systematically and thoroughly. No area should be omitted. Every room, closet, attic, concealed space, basement, areaway, or other out-of-the-way place where fire could start should be inspected. Note should be made of secret areas or processes from which the inspector is barred.

OBSERVATIONS

What should the inspector look for? The next few paragraphs give a general indication of what the inspector should be observing while going through the property. More specific information is contained in subsequent chapters.

Exterior

While touring the exterior areas of the property, the inspector should note the location and character of potential exposures and the condition of outdoor storage. Accessibility is an important factor. Fire lanes must be of sufficient width to allow the passage of fire apparatus. Hydrants and other sources of water must be accessible. Sprinkler valves must be open; standpipe connections must be capped, free of debris, and accessible. Topography is important. If there is a flammable liquid spill, in which direction will the product flow? What sort of drainage facilities are provided? Attention should also be given to dikes surrounding flammable liquids storage.

Building Construction

The type of construction and the materials used will influence the ease of ignition and the rate of fire spread. Openings in fire-rated walls need to be protected to retard or prevent the spread of fire. Doors to smokeproof towers must be kept closed to ensure a reasonably safe avenue of escape for occupants. The integrity of fire resistive walls and floor/ceiling assemblies must be assured. If holes are drilled in these assemblies for the passage of services and utilities and the voids are not sealed, they may allow the horizontal and vertical spread of fire. Many of these items are not apparent on a superficial basis, and it is often necessary to examine concealed spaces, such as ceiling voids, interiors of shafts, and stair enclosures.

Building Facilities

Such things as water distribution systems, heating systems, air conditioning and ventilating systems, electrical distribution systems, refuse handling equipment, and conveyor systems play an important role in the fire hazard potential of the premises. They must be properly installed, used, and maintained in order to minimize the hazard. While fire inspectors are not responsible for maintaining such systems, they should be able to determine that the equipment is being properly used and maintained. This may necessitate the review of someone else's maintenance records as part of the inspection process.

Hazards

Control of the hazards of materials depends on their proper storage, handling, use, and disposal. In this regard, the inspector should pay particular attention to housekeeping and storage practices. The inspector should also be familiar with any process hazards or special hazards that may be present on the premises.

Fire Detection and Alarm Systems

Often a property will be equipped with various fire detection and alarm devices and systems. The purpose of such equipment may be to detect the presence of fire, alert the occupants, notify the fire department, or combinations of these functions. The fire inspector should understand the function of, and be able to identify, the major components of these systems. Routine inspections should ensure that manually operated stations or devices are clearly marked and accessible to occupants. Periodic tests by, or in the company of, the inspector should be performed to confirm that the systems are in operating condition. Detection, alarm, and signaling systems are addressed in Chapter 26.

Fire Suppression Equipment

Also coming under the watchful eye of the fire inspector is the fire suppression equipment provided on the premises. Typical equipment includes sprinkler and standpipe systems and portable fire extinguishers. Routine inspections will determine that sprinkler valves are open, sprinklers are not obstructed, and interior hose lines are in place. Inspectors also determine that portable fire extinguishers of the proper type for the hazard present are provided, serviceable, clearly identified, and accessible to the occupants.

Special extinguishing systems for special hazards are also subject to the inspector's attention.

The fire inspector will conduct or witness periodic operational tests of fire extinguishing equipment. These are discussed in greater detail in Chapters 28, 29 and 30

CLOSING INTERVIEW

At the conclusion of the tour of the facility, the inspector should discuss the results with a person in authority. The inspector may have found conditions which seriously jeopardize the safety of the occupants and the property itself and which should be corrected immediately. In-house fire inspectors or their supervisors often have the authority to remedy hazardous situations. The visiting inspector, however, may have to rely on persuasion to convince the owner's representative that corrective action should be taken at once.

For less urgent conditions, the inspector's recommendations should be clearly explained so that management fully understands why they are made. The inspector's viewpoint should be expressed in easy to understand terms without engaging in arguments, technicalities, or petty fault-finding — any of which will only antagonize the people the inspector most wants to influence.

REPORTS

Some form of written report should be prepared for each inspection. The amount of detail required will depend upon the character and purpose of the inspection. In general, every report should include the following items:

(a) Date of inspection;

(b) Name of inspector;

(c) Name and address of property with a notation of the name and title of the person interviewed;

(d) Name and address of owner (or agent if a different location);

(e) Names of tenants of a multiple occupancy building (but not necessarily including, for example, the name of every tenant in an apartment building or office building);

(f) Class of occupancy (If mixed occupancy, state each principal occupancy and its location. In the case of industrial plants, state the principal items of raw materials and finished product.);

(g) Dimensions of buildings including heights and type of construction (See Chapter 3.);

(h) Factors contributing to fire spread inside buildings (stairways, elevator and utility shafts, lack of vertical and horizontal cutoffs);

(i) Common fire hazards (open flames and heaters, friction, electricity);

(j) Special fire hazards (hazardous materials, their storage, handling, use, and processes);

(k) Extinguishing, detection, and alarm equipment;

(l) Employee firesafety organization;

(m) Exits (adequacy and accessibility);

(n) Exposures, including factors making fire spread possible between buildings; and

(o) Recommendations or notations of violations.

The purpose of the report is to paint an accurate picture of the property, its hazards, and its fire protection without going into unnecessary detail. An inspection report should give the reader a clear understanding of the conditions found and the corrections needed.

The exclusive use of the checklist form of report could easily lead to slipshod inspections. Where a procedure is routine, such as determining that a sprinkler valve is open, a checkoff may be adequate. However, where a measurement, such as water or air pressure, is to be checked, provision should be made for entering the actual measurement.

Hazardous practices and conditions are best treated in the narrative form of report. The inspector who is required to describe the conditions observed is likely to do a more thorough inspection job than one who merely completes a checklist. A checklist cannot be devised to take into account every situation that conceivably could arise. The use of standardized reporting terms is desirable; NFPA 901, Uniform Coding for Fire Protection, can be helpful in this regard.

The inspector's recommendations for reducing hazards and improving protection constitute an important part of the reporting process. Recommendations may be prepared as a separate document and submitted to the property owner or manager for consideration. A copy of the recommendations should be filed with the inspection report if they are not an integral part of the report.

SURVEYING AND MAPPING

During the initial inspection, the inspector should gather information that will be used to prepare a site plan if one does not already exist. Such information will include construction features, occupancy data, fire protection features, and exposures. The site plan is a scaled drawing that indicates the locations and dimensions of the buildings, fire protection equipment (including water distribution systems), and the specific hazards and hazardous processes in each building. In order to show details of fire protection features, it may be necessary to draw a series of side sketches, which need not be to scale.

ITEMS FOR DAILY INSPECTION

(Similar considerations apply to matters for which inspections are appropriate at other periodic intervals.)

A list of items that will need daily attention should be prepared for each property. This list will be different in different properties. To show the property owner and inspector the sort of things to be included in such a list, the following is an example taken from a list in actual use:

(a) Height of water in gravity tank and reservoirs;

(b) Check on pilot light and meters of fire pump control panels to be certain equipment is energized;

(c) Test of rust protection equipment on gravity tank to be certain that it is energized;

(d) Check of pilot lights and trouble lights on sprinkler

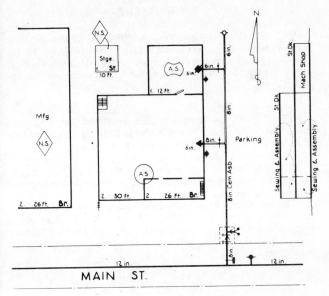

Figure 2-2. A typical site plan.

water flow alarm panels; and

(e) Reading of air pressure on dry pipe sprinkler valves.

In addition, there may be matters in the various departments of the plant requiring a daily check. A convenient routine is to provide a card for each item to be checked. Each card should be kept at the location to be checked, and departmental or other employees should be required to make and record the necessary daily observations. Card entries should show date, time, and by whom the observation was made.

It is not enough for management to specify daily checks. The individuals assigned to the checks must feel that if the matter is important enough to be recorded, it must be done correctly. The card records should be reviewed weekly by the property manager or the fire inspector, and the records summarized in the weekly report of loss prevention activities.

BIBLIOGRAPHY

McKinnon, G.P. ed., *Fire Protection Handbook*, 15th ed., National Fire Protection Association, Quincy, MA, 1981. Section 20, Chapter 1 contains the specific information needed to make a survey, as well as mapping symbols and common abbreviations used in mapping.

NFPA Codes, Standards, and Recommended Practices. (See the latest *NFPA Codes and Standards Catalog* for availability of current editions of the following documents.)

NFPA 174, *Standard for Fire Protection Symbols for Risk Analysis Diagrams*. Uniform fire protection symbols for use in property and structure diagrams for fire loss or risk analysis.

NFPA 901, *Uniform Coding for Fire Protection*, National Fire Protection Association, Quincy, MA. Part II of this document contains a glossary of terms that the inspector can use to provide standardized terms in inspection reports. There are additional terms for occupancy, processes, and equipment.

Chapter 3

BUILDING CONSTRUCTION

The type of construction and the materials used influence the life safety and fire protection requirements of a building. Inspectors have a major responsibility in determining that those requirements are met at all times. In order to discharge that responsibility, they must know the functions of various structural elements and understand the significant characteristics of construction types.

STRUCTURAL ELEMENTS

In general, the structural components of a building can be divided into two groups — those elements that support the structure (framing members) and those that enclose the working, storage, and living spaces (walls, floors, ceilings, and roofs).

Framing Members

The framing members form the skeleton of the building, which supports the building itself (dead load) as well as its contents, occupants, wind, snow, and ice (live loads). Common framing members include beams, joists, girders, rafters, and purlins, which are considered horizontal supports. Columns and load-bearing walls are vertical supports

Beams are horizontal structural members that are usually supported at both ends and may be supported by columns at one or more intermediate points. Joists are small, parallel beams that run from one wall to the opposite wall and support a floor or ceiling. If the span between walls is too great for continuous joists to provide adequate support, a beam may be used to support one end of each joist, while the other is supported at one of the opposing walls. If two or more beams are used, some joists will be supported at both ends by beams.

Girders are horizontal members that support vertical loads, the load exerted by an interior bearing wall, for example. A girder may consist of a single member or of several members bound together to act as a unit.

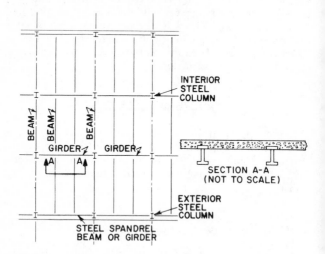

Figure 3-1. A partial floor plan of a steel frame structure.

Rafters are parallel beams that support a roof, and purlins are horizontal members in a roof assembly that support the rafters.

Vertical supporting members include columns and bearing walls. Columns often extend from the foundation to the roof and support the floors between. In modern construction, they are usually made of steel or reinforced concrete; in older structures one may find columns made of cast iron, timber, or stone. Bearing walls serve a dual purpose — vertical support and enclosing the building or spaces within the building.

Walls

Walls may be described in several ways — by function, application, or construction. Functionally, a wall is either load bearing (supports a load in addition to its own weight) or non-load bearing (supports only its own weight).

The following are examples of walls described by application. An exterior wall separates the interior from the exterior of a building. In a sense, a party wall is an exterior wall, because it separates the interior of the building from the outdoors. A party wall, however, is common to two buildings and has no exterior face. An interior wall of one story or less that separates two areas is known as a partition. Exterior walls, party walls, and partitions may be load bearing or non-load bearing.

Curtain walls are exterior walls that are usually supported by the structural frame. An enclosure wall is an interior wall that encloses an elevator shaft, stairwell, or other vertical opening.

A fire wall is a wall erected to prevent the spread of fire. Customarily, it is self-supporting and designed to maintain its structural integrity even in the event of the collapse of the structure on either side of the wall. A parapet wall is that portion of the wall extending beyond the roof.

Some walls are described by the type of construction used to make them. For example, a cavity wall consists of inner and outer masonry walls (sometimes called wythes) separated by an air space. The wythes are held together with metal ties. Hollow walls are of similar construction but contain no metal ties to hold the wythes together. Faced walls are composed of two different masonry materials that are bonded together to act as a unit under load. A veneered wall consists of a masonry facing attached to a backing; the two materials are not bonded together to act as a unit under load.

Sandwich panels are precisely what the name implies — sandwiches. An insulating core material is enclosed by two thin faces. The core is usually made of a foamed plastic, low density fiberboard, or a fibrous glass material. Exterior faces may be metal, cement asbestos board, or some similar siding material. The interior face may or may not be a different material.

Floor/Ceiling Assemblies

Although floor/ceiling assemblies do not bear the dead load of the building, they do bear the live load of the occupants, equipment, and machinery, and they do influence firesafety in buildings. Ceilings may be attached directly to the underside of the floor supports, or they may be suspended. The spaces above ceilings may contain combustible materials and may be used as part of the air handling system.

Ceiling materials that contribute to the fire resistance of a structure include lath and plaster, gypsum wallboard with finished joints, and certain mineral tile acoustical panels used in suspended ceilings.

Neither combustible ceilings nor ceilings that may fall when exposed to fire provide significant fire protection. Suspended ceilings that permit the passage of hot gases and flame into the space between the ceiling and the floor or roof supports above are ineffective.

Roofs

Roofing systems are constructed with several combinations of materials and in a variety of configurations. Basically, a roof consists of supports, a deck, and a covering.

Supports may be steel beams and open web steel joists, reinforced concrete beams, wood rafters, or steel or wood trusses.

The roof deck, along with its supporting members, carries the live load of wind and snow and any roof-mounted machinery. It also acts as the base for roof coverings, which provide insulation and protection against the weather. Typical roof deck materials are wood, reinforced concrete, and steel. Concrete decks may be fabricated with precast panels or poured in place over a light steel form.

Roof coverings are categorized as prepared or built-up. Examples of prepared coverings are fire retardant treated wood shingles, brick, concrete, tile, and slate. Untreated wood shingles have been used in the past, but are now prohibited in many communities because they ignite readily and are capable of producing flying brands, which may serve as ignition sources for other combustibles. A built-up roof consists of several layers of materials. For instance, tar and gravel roofs are made with several layers of roofing felt and insulating panels bonded together with an adhesive (often hot tar) and topped with roofing gravel. A vapor seal may also be used between the deck and the insulation.

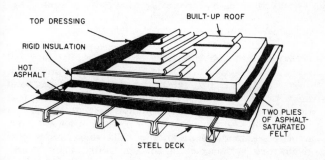

Figure 3-2. A typical built-up roof covering.

TYPES OF CONSTRUCTION

A classification system defines four standard types of construction — noncombustible/limited-combustible, heavy timber, ordinary, and wood frame. At one time, the term "fire resistive" was used to describe a construction type. Buildings traditionally classified fire resistive are now included in the noncombustible/limited-combustible category.

Noncombustible/Limited-Combustible Construction

Buildings classified as noncombustible/limited-combustible construction are made with materials that do not contribute to the development or spread of fire. Typical of this class of construction are metal-framed, metal-clad buildings and concrete block buildings having metal deck roofs supported by unprotected open web steel joists.

The principal hazard of such structures is their potential for collapse when exposed to fire. Therefore, noncombustible/limited-combustible construction should be used where the anticipated fire severity is low or where the available fire suppression agents and equipment are sufficient to deal successfully with the most severe fire that could be expected to occur.

Heavy Timber Construction

To qualify as heavy timber construction, a building must meet several requirements. Bearing walls and bearing portions of walls must be of noncombustible materials, have a fire resistance rating of 2 hours or more, and remain stable under exposure to fire. All exterior walls, whether bearing or nonbearing, must be made of noncombustible materials. Columns, beams, girders, and roof deck planks must have specified minimum dimensions in order to qualify as heavy timber construction. Floors and roofs are built of wood, generally without concealed spaces. However, where concealed spaces are permitted, they must be closed with tightly fitting wooden covers. Structural components of materials other than wood of specified sizes must have a fire resistance rating of at least 1 hour. Typical of heavy timber construction in the United States are the textile mills built in New England in the nineteenth century.

Glued laminated timbers are used in modern heavy timber construction. Such timbers that conform to the specifications of the National Forest Products Association have performed in a manner similar to solid sawn beams of a similar size.

A desirable characteristic of heavy timber construction is that it provides a slow burning fire in comparison to other forms and materials of construction. The surface areas of structural members are small in relation to their volume, and heat conduction through them is relatively slow. The surface char that develops on the burning timbers acts as insulation and slows heat penetration into the timbers.

Ordinary Construction

Ordinary construction at one time was a popular construction method for mercantile buildings, schools, churches, hotels, and institutional occupancies. In ordinary construction, exterior bearing walls and bearing portions of exterior walls are of noncombustible or limited-combustible materials (such as brick, concrete, or reinforced concrete), carry hourly fire resistance ratings, and exhibit stability when exposed to fire. Nonbearing exterior walls are also made of noncombustible or limited-combustible materials. Floors, roofs, and interior framing are made entirely or in part of wood, but in smaller dimensions than are required for heavy timber construction.

Floor joists in ordinary construction are much closer together than are the floor beams in heavy timber construction, thus creating more channels beneath the floors. Sheathing wall and floor framing with gypsum wallboard, for example, will provide some fire protection for those framing members. At the same time, the sheathing creates concealed wall and ceiling spaces in which fire may originate or through which fire may spread. Firestopping then becomes an important factor in the building's ability to resist the spread of fire. In industrial occupancies or other buildings where appearance is not a factor, open joist construction may be used. Although the absence of sheathing increases the area of exposed wood, fighting fire in joist channels then becomes easier, provided the involved channels are in an area accessible to fire fighters.

When roofs and floors and their supports have fire

resistance ratings, ordinary construction may be designated "protected ordinary construction."

Wood Frame Construction

Construction having exterior walls principally or entirely made of wood or other combustible material and not qualifying as heavy timber or ordinary construction is known as wood frame or simply frame construction.

Typically, walls and partitions are framed with 2-in. by 4-in. wood studs attached to wood sills and plates. Wood boards, plywood sheets, or various composition boards are nailed to the studs. Over this underlayment is placed a layer of building paper and then the finishing material. The exterior wall covering may be any one of a variety of materials including wood or cement-asbestos shingles; clapboards of wood, plastic, or metal; matched boards; brick veneer; sheet metal cladding; stucco; and cement-asbestos corrugated sheets.

In wood frame construction, firestopping within walls and partitions and between floors is important. Perhaps the most common method of firestopping walls is by nailing snugly fitting pieces of 2-in. by 4-in. pieces of lumber between the studs. Some old buildings were constructed with bricks laid solidly in the spaces between wood studs. These were known as brick-nogged walls. Another acceptable practice is filling the voids and concealed spaces with noncombustible insulating materials.

BIBLIOGRAPHY

Brannigan, Francis L., *Building Construction for the Fire Service*, 2nd ed., National Fire Protection Association, Quincy, MA, 1982. In Chapter 2, Principles of Construction, the author discusses the hazards of trusses when exposed to fire.

Building Materials Directory, Underwriters Laboratories Inc., Northbrook, IL, issued annually. This directory is a listing of building materials and hardware that have been subjected to fire tests and, as a result, have been found suitable for use in building construction.

Fire Resistance Directory, Underwriters Laboratories Inc., Northbrook, IL, issued annually. This directory contains hourly fire resistance ratings for beams and columns as well as floor, roof, wall, and partition assemblies that have been subjected to standard fire tests.

McKinnon, G.P. ed., *Fire Protection Handbook*, 15th ed., National Fire Protection Association, Quincy, MA, 1981. Chapter 4 is a detailed description of the standard types of building construction. Chapter 5 discusses structural framing elements and describes the various floor, wall, and roof assemblies used in building construction.

Chapter 4

PROTECTION OF OPENINGS

One method of limiting the spread of fire in a structure is to divide the interior into compartments by means of fire barriers — fire walls and rated floor/ceiling assemblies. Fire barriers can be expected to delay the spread of fire from the room or area of origin to other parts of the structure only if they are properly constructed and maintained and if their openings are protected.

THE HAZARD

It is not uncommon for heated unburned pyrolysis products to flow out of the area of initial involvement, mix with air, and ignite. Such flame extension can occur over noncombustible surfaces. Flames may heat interior finish materials to the point where they release pyrolysis products of their own, which also ignite and contribute to the intensity of the extending flame. Properly maintained opening protection is essential to containing the fire until suppression activities have been initiated. Part of your responsibility, therefore, is to determine that nothing has been done to nullify the protection of openings.

FORM OF PROTECTION

A variety of methods for protecting openings in fire barriers is available. The method selected will depend on the type and function of the opening. Typical protection measures include firestopping, fire resistive construction, fire doors, and wired glass. Special problems will require other forms of protection.

VERTICAL OPENINGS

Unprotected openings in floors and ceilings — vertical openings — permit the vertical extension of fire from one floor to another.

Floor/Ceiling Penetrations

Where holes are made through floor/ceiling assemblies for the routing of cables, conduits, or pipes, air gaps are created which permit the passage of combustion products from floor to floor if the gaps are left unsealed. One method used to seal these gaps involves modular devices sized for the pipe, conduit, or cable, which contain an organic compound that expands when heated and seals the penetration. Another employs a foamed-in-place fire resistant silicone elastomer that expands as it foams and seals the penetration.

The penetrating objects should be supported sufficiently to prevent placing a mechanical stress on the seal that may pull the sealant from the opening. Where "temporary" routing of utilities or control cables is a fairly common occurrence, workers tend to neglect sealing the gaps. Be alert for such conditions. Often utility lines are housed in closets and, therefore, are not obvious. Learn the locations of these concealed, but accessible, fire barrier penetrations, so that you do not overlook them in the inspection process.

Stairways, Shafts, and Chutes

Certain vertical openings cannot be sealed because their functions require that they communicate between floors. Examples include stairways, elevator shafts, utility shafts, and chutes for packages, laundry, or trash. Such openings are enclosed in fire resistive shaft construction. Openings in the walls of stairwells and elevator shafts should be protected by rated self-closing fire doors. People have been known to prop open stairway doors for the sake of convenience. This, of course, defeats the purpose of a fire door and is a condition that must be corrected immediately. Openings in the walls of utility shafts are protected with self-closing fire doors or access doors approved or listed for the purpose.

Escalators

Holes made in floor/ceiling assemblies to accommodate escalators present a unique protection problem because it is not practical to enclose them in fire resistive construction. There are, however, alternative forms of protection.

One method relies upon a combination of automatic fire or smoke detection, an automatic exhaust system, and an automatic water curtain. A second method fills the opening with a dense water spray pattern from open, high velocity water spray nozzles. The water spray system is operated automatically by heat or smoke detection and is equipped with manual control valves to minimize water damage. As is the case with other water spray and sprinkler systems, you should examine the control valves to make sure that they are open.

A self-closing rolling shutter, actuated by automatic heat or smoke detection, can be used to enclose the top of the escalator. However, this method is not recommended for use between the basement and ground levels.

Another method is to protect the opening with a partial enclosure of fire resistive construction in a "kiosk" configuration. The enclosure is equipped with self-closing doors. Check that the doors are in operating condition and that the self-closing feature has not been circumvented in any way.

HORIZONTAL OPENINGS

Openings in fire walls and partitions — horizontal openings — if left unprotected, will aid the spread of fire in the horizontal plane; that is, throughout the floor of origin. Corridors must be protected not only because they are a vehicle for the horizontal spread of fire, but also because they are a part of the means of egress for occupants.

Fire Doors

One of the most widely used means of protecting openings in fire resistive walls is the fire door. Suitability of a door assembly for use as a fire door is determined by tests conducted by independent testing laboratories. Fire doors may be given an hourly rating, an alphabetical letter designation (Class A, B, C, D, or E), or a combination of the two. Current practice, however, is to specify a fire door by its hourly fire resistance rating.

Ratings: Each classification of fire door has specific applications.* Where a wall separates two buildings or divides a building into two fire areas, a 3-hour fire door is used. Openings in walls enclosing hazardous areas may also be protected with 3-hour doors.

Doors in openings in exterior walls that may be subjected to severe fire exposure from outside the building and doors protecting openings in 2-hour enclosures of vertical openings in buildings carry a 1½-hour fire resistance rating.

One-hour enclosures of vertical openings are protected by 1-hour fire doors.

Three-quarter-hour fire doors are used to protect openings in exterior walls of buildings that may be subjected to a light or moderate fire exposure from outside the building.

Fire doors having ½- and ⅓-hour ratings are intended primarily for smoke control. They are used across corridors where a smoke partition is required and in openings in partitions of up to 1-hour fire resistance rating between a habitable room and a corridor.

Construction: Several types of construction are used in the manufacture of fire doors.

Composite doors are flush doors made of a manufactured core material with chemically impregnated wood edge banding and faced with untreated wood veneer or laminated plastic, or encased in steel.

---

*In local codes or ordinances, the fire resistance rating required for a specific application may vary from those given here.

Hollow metal doors are made in flush and panel design of 20-gage or heavier steel.

Metal-clad or Kalamein doors are flush or panel design swinging doors of metal covered wood cores or stiles and rails and insulated panels covered with 24-gage steel or lighter.

Sheet metal doors are made in corrugated, flush, or panel designs of 22-gage steel or lighter.

Rolling steel doors are fabricated of interlocking steel slats or plate steel.

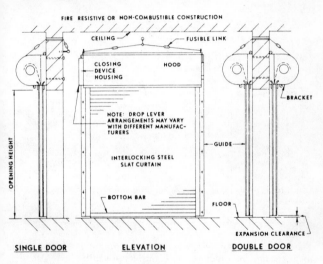

Figure 4-1. Surface mounted rolling steel doors (fusible links required on both sides of wall).

Tin-clad doors are of two- or three-ply wood core construction. They are covered with 30-gage galvanized steel or terne plate (maximum size, 14 by 20 in.) or with 24-gage galvanized steel sheets (maximum width, 48 in.).

Curtain type doors consist of interlocking steel blades or a continuous formed spring steel curtain installed in a steel frame.

Wood core doors consist of wood, hardboard, or plastic face sheets bonded to a wood block or wood particle board core material with untreated wood edges.

Door Closing: Fire doors must be self-closing or automatically closing in the event of fire. A suitable door holder/release device may be used provided the automatic release feature is actuated by a combination of automatic fire detection devices, such as smoke detectors or fusible links.

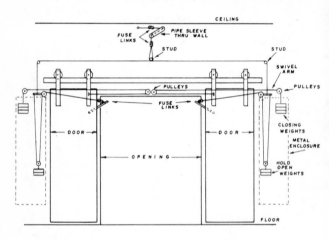

Figure 4-2. Closing devices for center parting horizontally sliding doors (fusible links required on both sides of wall).

Maintenance and Inspection: Fire doors, normally open during working hours, should be closed during nonworking hours.

Highly combustible material, which is likely to produce a flash fire, should not be stored near an opening in a fire wall, as fire might communicate through the opening before the protective device could operate.

Observe that fire doors are not obstructed or blocked in any manner or intentionally wedged open so that free self-closing is not possible. Substantial guards should be provided where it is necessary to prevent employees from piling stock against or near a door. In the event of intentional blocking or wedging, determine the reason for it and initiate appropriate action. Where doors have been fastened open to ensure increased ventilation, other ventilating means should be provided. Be sure that the movement of balance weights is free and unobstructed and that fusible links are of the proper temperature rating.

With tin-clad doors, there are some special points that require attention. Note that the door has proper lap over the opening. The binders are sometimes filled with blocking to make the door easier to open. These blocks should be removed. The door should have chafing strips, which add to the fire resistance of the door. Note the condition of the door. Is the tine covering corroded, torn, or battered? Is there evidence of dry rot? Edges caving in and screws pulling out may indicate dry rot. Tapping the door with a weighted object such as a hammer may provide some indication of the extent of the rot damage.

When inspecting fire doors, except rolling steel doors, perform an operating test to make sure that the doors do not jam or stick and that hardware is not incomplete or loose. Check the automatic closing mechanism by lifting the counterbalance weight or dropping the suspended weight. With door checks equipped with a fusible element, the test is limited to general observation of the device. Make sure that the fusible link in not corroded or coated with dirt or paint that would act as an insulator and reduce its temperature sensitivity. Rolling steel doors should be tested during nonworking hours so that, if a malfunction does occur, it will not interfere with normal activities on the premises.

Fire Shutters

Fire shutters are used to protect openings in exterior walls. If the potential fire exposure from outside the building is severe, 1½-hour shutters are used. Where the potential fire exposure is moderate or light, ¾-hour shutters are used. Shutters should be equipped to close automatically in case of fire. Do not overlook these devices on the inspection tour. If fire shutters are installed on the outside of the opening, they should be protected against the weather to ensure proper operation.

Wired Glass

Wired glass is used as the glazing material in ¾-hour fire resistance rated windows. Fire windows are designed for use in protecting openings in corridor and room partitions and in exterior walls where the potential exposure is moderate or light. Some fire windows are equipped with automatic closing devices that are actuated by automatic fire detection equipment. When inspecting fire windows, be sure that the closing devices are in operating condition.

Wired glass is frequently used as a vision panel in smoke stop barriers and in fire doors protecting stairway enclosures. While conducting your inspection, be alert for situations in which plain window glass may have been used to replace broken wired glass.

SPECIAL PROBLEMS

Duct and conveyor systems that penetrate walls and partitions as well as floors and ceilings contribute to both the horizontal and vertical spread of fire. Duct systems are discussed in Chapter 10, Air Conditioning and Ventilating Systems, and conveyors are discussed in Chapter 12, Materials Handling Systems.

BIBLIOGRAPHY

McKinnon, G. P. ed, *Fire Protection Handbook*, 15th ed., National Fire Protection Association, Quincy, MA 1981. Section 5, Chapter 9 discusses the effects of openings in fire barriers on the spread of fire in buildings and the protection of those openings.

NFPA Codes, Standards, and Recommended Practices. (See the latest *NFPA Codes and Standards Catalog* for the availability of current editions of the following documents.)

NFPA 80, *Standard for Fire Doors and Windows*. This standard describes the requirements for, and the installation and maintenance of, fire doors, windows, and shutters.

NFPA 80A, *Recommended Practice for Protection of Buildings from Exterior Fire Exposures*. One of the subjects addressed by this recommended practice is the protection of openings in exterior walls from exposure fires.

NFPA 101®, *Code for Safety to Life from Fire in Buildings and Structures*. This standard specifies the required hourly ratings for openings in fire rated barriers.

Chapter 5

INTERIOR FINISH

Interior finish is defined in the *Life Safety Code*® (NFPA *101*) as "the exposed interior surfaces of buildings including, but not limited to, fixed or movable walls and partitions, columns, and ceilings." Interior floor finish means the exposed floor surfaces of buildings including coverings that may be applied over a normal finished floor. Decorations and furnishings are generally considered apart from interior finish materials.

HAZARDS OF FINISH MATERIALS

The combustibility of interior finish materials has a major effect on the fire hazard of, and life safety in, buildings. Combustibility is a measure of the speed with which flame will spread along the surface of the material, the amount of smoke generated at the burning surface, and the amount of fuel the material contributes to the fire. A material having a high flame spread rating promotes the rapid spread of fire. One that has a high fuel contribution rating quickly aids in increasing fire intensity. A high smoke generation rating contributes to the life hazard in a burning building.

Flashover is the simultaneous ignition of all the combustibles in a room brought about by the contents being heated to their ignition temperatures, generally by radiated heat fed back from the ceiling and upper walls. A highly combustible finish material hastens the time from ignition to flashover due to its contribution to fire intensity over a wide area.

TYPES OF FINISH MATERIALS

Interior finish materials are chosen for their eye appeal,

acoustical and insulation values, durability, and ease of maintenance. Among the materials used are plaster, gypsum wallboard, wood, plywood paneling, fibrous ceiling tiles, plastics, and a variety of wall coverings.

Ordinary thicknesses of paint and one or two layers of wallpaper are not considered to have a significant effect on flame spread. However, heavy layers of these materials can alter the flame spread rating of the unfinished materials.

Plywoods and Veneers

Plywood and wood veneers consist of surface layers of thin wood glued to a backing, which, itself, may be several layers or solid wood. Where the glue used is heat resistant, the assembly has the burning properties of solid wood, but where it softens under heat, the thin outer layer delaminates and may burn with great rapidity. Douglas fir plywood of U.S. manufacture is generally made with heat resistant adhesives and thus burns like solid wood. However, there is no control over the properties of the adhesive, and you have no way of assuring yourself of the quality of the product, except in the case of tested and labeled plywood.

Acoustical Materials

Soundproofing effects can be obtained with draperies and thick rugs, which can be flame retardant treated where necessary. More commonly, acoustical materials are fixed to substantial portions of walls and ceilings, generally in the form of acoustical tile about 1-foot square, but sometimes in sheets covering a considerable area. Acoustical tile is usually about 1-in. thick and is characterized by perforations or irregular indentations, which are planned with relation to the wavelength of the sound they are designed to attenuate.

Acoustical materials vary from highly combustible fibrous construction to plaster or metal, which are obviously noncombustible. Often visual inspection gives no clue to flame spread properties. Acoustical materials can be coated with flame resistant paint without destroying their sound absorbing qualities, provided the perforations or indentations are not filled with paint.

Cemented Tiles

Tiles cemented to walls or ceilings will fall as soon as the cement is heated to its softening temperature. This does not affect flame spread where the tiles are completely noncombustible and where the space behind them is also noncombustible. However, where the tiles have a flame resistance treatment only on the exposed surface or where the backing to which the tiles are affixed is combustible, the effect of flame spread may be serious. A way of ensuring that tiles remain in place is through the use of metal fastenings, such as nails, screws, or clips.

Fire Retardant Coatings

The combustibility of any surface can be reduced by the application of a fire retardant coating or paint properly applied. Some fire retardant paints are intumescent; that is, under the influence of heat, they bubble to form a heat insulating coating. Some fire retardant paints or coatings deteriorate under the influence of moisture and may require periodic renewal.

Insulation

The use of noncombustible heat insulating materials in concealed spaces tends to reduce the hazard; combustible materials increase it. Many insulating materials, including noncombustible materials, such as mineral wool, are provided with vapor barriers often having a high flame spread rating. Where exposed to form the interior finish of attics and other spaces, vapor barriers of low combustibility are available.

Certain insulating materials are actually noncombustible, but many have binders, adhesives, or vapor barriers that are combustible. The hazard of insulation depends upon its combustibility, and materials having a high flame spread rating, such as cottons, are commonly treated to reduce combustibility, though the permanence of the treatment may be open to question.

Polystyrene and polyurethane insulations, usually in sandwich construction, are popular in building construction. The safe use of these materials is a function of the tightness of panel construction and the fire resistance of the outer surfaces of the panels.

Fires starting in, or extending to, combustible concealed spaces are difficult to extinguish and, when breaking out of concealed spaces, may quickly involve large areas. Concealed spaces are directly related to interior finish, because it is the interior finish that is the surface of the concealed space, at least on one side.

Concealed spaces vary from the channels between wooden studs closed off at top or bottom by firestopping, which are too small to involve any great fire danger, to open attic spaces and spaces above suspended ceilings, which may extend over an entire building. The degree of hazard is roughly proportional to the size of the space and the character of the materials exposed to fire therein.

APPLICATION METHODS

It is important that manufacturers' instructions for the application of interior finish materials be followed carefully, as improper application methods have a serious effect on the fire behavior of the finish material. Size and spacing of nails or other fasteners, type and application of adhesives, and the number of coats and application rates of fire retardant coatings are among the critical factors involved in the application of finish materials.

The substrate material to which an interior finish is applied will affect the fire hazard. A thin combustible finish applied to a noncombustible substrate may present little hazard. However, the hazard will be considerably greater if the same material is applied to a combustible base. For example, the hazard of combustible thin plywood veneer paneling is lessened considerably when it is applied to a gypsum wallboard substrate rather than directly to wood studs.

TESTS OF INTERIOR FINISHES

Building codes and the *Life Safety Code* specify the use of materials having certain fire ratings for interior finish. Most codes base requirements for interior wall and ceiling finish on a test known as the Steiner tunnel test. In the search for improved test methods over the years, a number of other tests have been developed.

Steiner Tunnel Test

The tunnel test is on a scale sufficient to measure, under conditions closely simulating actual fire, those factors of flame spread, smoke, and fuel, which are major items in loss of life by fire. Flame spread is the factor most generally considered, as the other factors are roughly proportional in the case of most materials. In the case of materials of unusual properties, the other factors may be of equal or greater importance. This test does not record the toxicity of the gases generated, but products of combustion can be identified.

The tunnel test also covers the factor of adhesion, if the sample is affixed to the top of the tunnel in the same manner as in actual use. Test results from material screwed or nailed to the surface with no air space may not apply to actual installations in which the material is glued on with a cement having a low melting temperature, or when the material is mounted on wood strips with space behind it in which combustibles are exposed.

Eight-foot Tunnel Furnace

In an effort to design a scaled down Steiner tunnel furnace, the Forest Products Laboratory developed an 8-foot tunnel test. Eventually, the design of the 8-foot tunnel furnace evolved into something quite different from the Steiner furnace. This test is used primarily to measure flame spread for research and product development purposes. However, some authorities will accept results of this test in lieu of other test data.

Radiant Panel Test

The National Bureau of Standards developed a radiant panel furnace test that utilizes a 6- by 8-inch test specimen and measures flame spread and rate of heat release. It is used for research and development purposes.

Corner Test

A fire test involving a 25-foot-high corner configuration with wing walls up to 50 feet long have been used to examine the burning characteristics of cellular plastics and the effectiveness of fire protection measures.

Small-Scale Tests

The American Society for Testing and Materials has adopted a number of small-scale tests for specific materials — treated wood, paints, treated paper and paperboard, textile floor coverings, and plastics. These tests are used primarily in product development and experimental laboratory work involving the flammability of building materials.

DECORATIONS AND FABRICS

You will frequently find situations where decorations or fabrics are a fire hazard. The hazard depends on the ease with which the materials are ignited by small sparks, cigarettes, electrical defects, and similar fire causes. If the decorations or fabrics are extensive, there is a life hazard due to the potential for a flash fire. Most localities, by law, prohibit combustible decorations or fabrics in places of assembly. A common requirement is that, if quick burning decorative materials are used, they must be treated with a flame retardant to reduce their combustibility. This requirement may apply to theater curtains, scenery, draperies, bunting, upholstery, and artificial plants, trees, and flowers, Christmas trees, paper decorations, and similar items. Flame retardant treatments are also necessary for tents and air-supported structures.

Noncombustible (glass) fabrics as well as a great variety of other treated fabrics that have varying degrees of reduced combustibility are available.

SI UNITS

The following factors are given as a convenience in converting to SI units the English units used in this chapter.

$$1 \text{ in.} = 25.4 \text{ mm}$$
$$1 \text{ ft} = 0.305 \text{ m}$$

BIBLIOGRAPHY

American Society for Testing and Materials, Philadelphia.

The following are ASTM-adopted small-scale tests of the flammability of building materials. These tests are used primarily for product development and experimental laboratory work.

Combustible Properties of Treated Wood by the Crib Test, ASTM E 160-50.

Combustible Properties of Treated Wood by the Fire Tube Apparatus, ASTM E 69-50.

Fire Retardancy of Paints (Cabinet Method), ASTM D 1360-58.

Flammability of Plastics 0.050 in. and Under in Thickness, ASTM D 568-77.

Flammability of Rigid Plastics Over 0.050 in. Thickness, ASTM D 635-77.

Flammability of Flexible Thin Plastic Sheeting, ASTM D 1433-77

Incandescence Resistance of Rigid Plastics, ASTM D 757-77

Flammability of Treated Paper and Paperboard, ASTM D 777-74

Flammability of Rigid Cellular Plastics, ASTM D 3014-76

Flammability of Plastics Using the Oxygen Index Method, ASTM D 2863-77.

Flammability of Finished Textile Floor Covering Materials, ASTM D 2859-76.

McKinnon, G. P. ed., *Fire Protection Handbook*, 15th ed., Quincy, MA 1981. Section 5, Chapter 6 deals with the types of interior finish, their application and role in building fires, and fire test methods.

NFPA Codes, Standards, and Recommended Practices. (See the latest *NFPA Codes and Standards Catalog* for the availability of current editions of the following documents.)

NFPA 255, *Method of Test of Surface Burning Characteristics of Building Materials*. This standard contains the specifications and procedures for testing building materials with the Steiner tunnel furnace.

NFPA 701, *Standard Methods of Fire Tests for Flame-Resistant Textiles and Films*. The specifications and procedures for testing the flame resistance of materials used in decorations, tents, and air-supported structures are given in this standard.

Chapter 6

CONSTRUCTION, ALTERATION, AND DEMOLITION OPERATIONS

As a fire inspector, you may be faced with the hazards introduced during construction, alteration, and demolition operations at one time or another. Buildings are usually more vulnerable to fire at these times than at any other. The amount of combustibles and hazardous materials present may be greater than under normal operating conditions. The number of potential ignition sources may be greater, and fire protection systems may be impaired or even inoperative.

CONSTRUCTION

Construction projects progress more rapidly at some stages than at others. You should step up the frequency of your inspections during those periods when conditions are changing quickly.

Site Preparation

For new construction projects, fire protection begins with site preparation. The area should be stripped of vegetation to provide an adequate clear space around the work area. Means of access to the job site should be provided for the passage of fire apparatus.

A water supply for fire fighting purposes should be available on the site. Ideally, the permanent water supply would be installed early and serve the purpose. However, there are times when it may be necessary to provide a temporary supply in the form of aboveground water mains. Where temporary water mains are used, they should be protected from damage by construction equipment.

Temporary Structures

Temporary structures are used on construction sites for offices and the storage of equipment, tools, and supplies. At one time, these structures were commonly of wood construction. More recently, however, converted mobile homes, trailers, or premanufactured buildings have been used. When made of combustible materials, temporary structures should be located 30 feet or more from the building under construction, and they should be detached from one another.

The hazards are similar whether a mobile structure or a temporary structure is used. Look for overloaded temporary wiring and improperly installed portable or temporary heating equipment. Temporary utility lines and fuel storage facilities should be protected from damage by vehicles and construction equipment.

Temporary enclosures of fabric or plastic materials are often used to protect workers and construction operations from the weather until the building is enclosed. If a fabric is used, be sure that it is a fire-retardant-treated tarpaulin. If plastic is used, it should be a slow burning type. In either case, the material should be fastened securely to a rigid wood or steel frame to prevent it from coming in contact with an ignition source. If you see a tarpaulin or plastic sheet flapping in the breeze, it is a sure indication that the enclosure is not properly anchored.

Precautions

As construction operations progress, rubbish and trash accumulations can become a problem. Such materials should be collected and disposed of daily. If it is necessary to store debris at all, suitable containers should be provided. The containers should either be a safe distance from the building and temporary structures, or be protected to prevent exposing the building to fires in them.

Internal combustion engine driven machinery, such as pumps, air compressors, and hoists, is used on building construction sites. Be sure that such equipment is placed so that the exhaust discharge is directed away from combustible materials. If the exhaust is piped outside the building, be sure that there are at least 6 inches of clearance between the pipe and any combustibles. Do not permit refueling operations to take place while the engines are running. Do not allow bulk storage of fuel inside or near the building.

Open flame devices are an obvious source of ignition. Therefore, their use must be strictly controlled. This is particularly true of cutting and welding operations in which bits of burning metal may travel relatively long distances and still have enough heat energy to ignite combustibles. The procedures and precautions discussed in Chapter 24, Welding and Cutting, should be employed.

Cold weather operations call for the use of portable heaters (salamanders). Make sure that their location and use is carefully controlled and that the area is kept free of combustibles.

Flammable liquids, principally in the form of fuels and solvents, are an obvious hazard found on construction sites. Small quantities can be safely used on site; however, be sure that they are stored in, and dispensed from, approved safety containers. Storage of flammable liquids in bulk quantities should be avoided, but if bulk quantities are need, they should be stored in an approved manner, and all local permits must be obtained. Guidance on the storage of flammable liquids is given in Chapter 15 and in NFPA 395, *Standard for the Storage of Flammable and Combustible Liquids on Farms and Isolated Construction Projects.*

Asphalt and tar kettles are another common hazard associated with building construction. See that the kettles are equipped with covers that can be used to smother flames in the event a fire does occur. Make sure that dry chemical, not water type, fire extinguishers are provided for the protection of the kettles. The use of water on burning tar will cause the material to froth and possibly spill over, spreading the fire. Roofing mops soaked with tar have been known to ignite spontaneously and cause fires. Do not allow used mops to be left indoors or near ignition sources or combustible materials. The mops should be cleaned thoroughly and stored carefully.

Fire Protection

In order to provide some fire fighting capability during construction, standpipes, where required in the finished structure, should be installed on a floor-by-floor basis. Until protection is complete, the system is usually a dry-pipe system equipped with a fire department siamese hose connection. On your inspection tours, make certain that the siamese connections are accessible to the fire department. Do not allow storage or parking near them. In the absence of a standpipe system, free access to some form of temporary or portable first aid fire protection equipment should be provided.

ALTERATIONS

More and more older buildings are being altered or renovated, some to preserve the architecture of an age gone by. As older buildings are rehabilitated, every effort is made to bring them into compliance with present-day building and fire codes. This is not always possible. In such cases, equivalent protective measures may be acceptable in lieu of meeting certain code requirements. The equivalencies may involve the installation of automatic sprinkler protection, smoke detection systems, and smoke control features. Your responsibility as a fire inspector is to police those protection systems as you would in a new building.

Sometimes renovations have to be made while the building is partially occupied. It is extremely important to life safety that exits for occupants are properly maintained. It is your job to make sure that they are — that they are accessible and properly identified. In addition to exits for occupants, at least one exit should be provided for construction workers in the area under renovation. You should be just as diligent in inspecting this exit as you are in inspecting the public exits.

If a building addition blocks an existing exit, an alternate means of egress must be provided. You must inspect that alternate route as part of your normal procedure. Is the exit free of debris or storage? Is it properly identified and lighted? Is the exit discharge clear of parked vehicles or other objects?

DEMOLITION

Demolition operations have many of the attendant hazards of construction operations as well as a few others. The hazards of cutting torches, flammable liquids, and trash accumulations are present in demolition operations as they are in construction operations.

Early in a demolition project, flammable liquids and combustible oils should be drained from tanks and machinery and removed from the building immediately. The removal of residue and sludge deposits is important, especially in areas where cutting torches are used.

Fixed fire protection systems should be maintained for as long as possible. Sprinkler and standpipe systems should be modified so that they can be dismantled floor by floor as demolition progresses downward from the top floor. This will preserve protection on the floors below. Either type of protection can be readily converted from a wet-pipe system to a dry-pipe system.

Generally, chutes are provided to carry demolition rubble and debris from the upper floors to trucks or mobile trash receptacles below. They should be erected outside the building. The use of inside chutes would necessitate cutting holes in the floor, thereby creating an unprotected vertical opening through which fire can spread rapidly from floor to floor.

SI UNITS

The following conversion factors are given as a convenience in converting to SI units the English units used in this chapter.

$$1 \text{ in.} = 25.4 \text{ mm}$$
$$1 \text{ ft} = 0.305 \text{ m}$$

BIBLIOGRAPHY

"Collapse of the Hotel Vendome, Boston, Mass.," *Fire Journal*, Vol. 67, No. 1 (Jan. 1973), pp. 33-41. This is an account of the collapse of an old Boston hotel during rehabilitation operations.

Herbstman, Donald, "Fire Protection During Construction," *Fire Journal*, Vol. 64, No. 1 (Jan. 1970), pp. 29-32, 89. This is a description of the procedures, practices, and equipment used to protect the World Trade Center in New York during its construction.

McKinnon, G. P. ed., *Fire Protection Handbook*, 15th ed., National Fire Protection Association, Quincy, MA, 1981. Section 5, Chapter 14 discusses the fire hazards of construction, alteration, and demolition of buildings.

Sharry, John A., "Group Fire Indianapolis, Indiana," *Fire Journal*, Vol. 68, No. 4 (July 1974), pp. 13-16. Fire originating on the top floor quickly spread throughout a partially demolished building and involved four other buildings.

NFPA Codes, Standards, and Recommended Practices. (See the latest *NFPA Codes and Standards Catalog* for the availability of current editions of the following documents.)

NFPA 51, *Standard for the Design and Installation of Oxygen-Fuel Gas Systems for Welding and Cutting*. A standard for acetylene, hydrogen, natural gas, LP-Gas, MAP, and other stable gases.

NFPA 51B, *Standard for Fire Prevention in Use of Cutting and Welding Processes*. A standard for the management of fire prevention and protection precautions.

NFPA 241, *Standard for Safeguarding Building Construction and Demolition Operations*. A standard on firesafety during the erection, alteration, and demolition of buildings.

NFPA 395, *Standard for the Storage of Flammable and Combustible Liquids on Farms and Isolated Construction Projects*. This standard addresses containers of 60 gallons or less capacity and tanks from 61 to 1,000 gallons capacity for flammable liquids storage.

NFPA 495, *Code for the Manufacture, Transportation, Storage, and Use of Explosive Materials*. Provides guidance on the storage and use of explosives, which will be helpful to workers in the demolition and construction industry.

Chapter 7

LIFE SAFETY CONSIDERATIONS

The principles of life safety in buildings are to provide sufficient exits to permit the prompt evacuation of occupants during a fire and to provide the means for preventing the occupants from being exposed to undue danger from the fire and the products of combustion for a period of time that can be considered reasonably necessary for escape in case of fire.

Life safety does not rely on any single safeguard for the preservation of life during a fire in a structure. It considers a number of elements. Part of the life safety system includes structural considerations, such as the subdivision of floors in a structure to provide areas of refuge, and vertical opening protection to prevent the spread of smoke and fire from one floor to another (see Chapter 4). Interior finish materials have a very significant effect on the speed with which flames will spread along the surface (see Chapter 5). Detection and signaling systems are vital to providing early warning to the occupants and automatically notifying the fire department (see Chapter 26). Fires in areas of unusual hazard may be capable of endangering occupants as they attempt to evacuate the structure. These areas should be safeguarded with special extinguishing equipment (see Chapters 28 and 29). The foregoing features all have a bearing on the amount of time that would be available for occupants to make a safe escape. Your role as an inspector in each of the areas is discussed in the chapters referenced here. The final parts of the system deal with the means of egress. Life safety depends on all components of the system being maintained, and regular and thorough inspections are an equally important part of the system. Such inspections should be frequent and, apart from ensuring that all requirements are met, exit routes and exit ways must be maintained free and clear and properly indicated at all times.

DEFINITIONS

A means of egress is a continuous and unobstructed way of exit travel from any point in a building or structure to a public way and consists of three separate and distinct parts: (a) the way of exit access, (b) the exit, and (c) the way of exit discharge. A means of egress comprises the vertical and horizontal ways of travel and includes intervening room spaces, doorways, hallways, corridors, passageways, balconies, ramps and stairs, enclosures, lobbies, escalators, horizontal exits, courts, and yards.

Exit access is that portion of the means of egress which leads to an entrance to an exit.

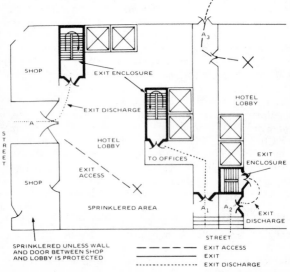

Figure 7-1. Examples of exit discharge. To the occupant of the building at the discharge level, the doors at A, A₁, A₂, and A₃ are exits and the path denoted by dashes is the exit access. To the person emerging from the exit enclosure, the same doors and paths denoted by dotted lines are the exit discharge.

An exit is that portion of a means of egress which is separated from other spaces of the building or structure by walls, floors, doors or by any other means to provide a protected way of travel to the exit discharge.

The exit discharge is that portion of a means of egress between the termination of an exit and a public way.

EXIT ARRANGEMENT

There are a number of acceptable exit arrangements and facilities with which the fire inspector should be familiar.

Doors

Doors used for exit purposes are side-hinged and swing in the direction of exit travel. Do not confuse exit doors with fire doors, which may swing, slide, or roll closed. Exit doors, such as those giving access to stairways, may also serve as fire doors. These are equipped with self-closers and may be equipped with automatic hold-open/release devices, which normally hold open doors and automatically release them in the event of smoke or heat detection or activation of an automatic sprinkler system. In the absence of automatic hold-open/release devices, watch for situations in which someone has used a wedge or hook to hold the door open. It is extremely important that doors in stairway enclosures be checked for proper operation and fit. Normally, doors to exit stairways must be closed; however, in certain low hazard occupancies, automatic hold-open/release devices may be permitted. Learn what the regulations are in your locale.

Panic Hardware

In places of public assembly, a room or area designed to accommodate fifty or more persons, and educational occupancies, exit doors are equipped with panic hardware. The releasing mechanism of such hardware is designed to yield to moderate pressure to allow the safe egress of occupants. One of the most flagrant violations of the life safety principles is the sealing of the hardware with a lock and chain on the excuse that such action is necessary for security purposes. It is not necessary. There are suitable devices available that will permit egress of the occupant while preventing the entrance of unauthorized persons.

Horizontal Exits

Horizontal exits provide for the movement of occupants from one building to an area of refuge in another at approximately the same level. Within one building, a way through or around a fire wall or partition to an area of refuge on approximately the same level is called a horizontal exit. Such arrangements are designed to provide sufficient refuge space for the people entering the area. Your job is to see that no one minimizes that space by using it for storage purposes and the integrity of the fire separation is maintained.

Stairs

Exit stairs should be arranged to minimize the danger of falling, as one person falling on a stairway can result in the blocking of an exit. Stairs should be wide enough so that two people can descend side by side; thus, a reasonable rate of evacuation can be maintained, even though aged or infirmed persons may slow the travel on one side. There should be no sudden decrease in width along the path of travel; this may create congestion or, in a panic rush, a solid wedge of bodies blocking the exit.

Steep stairs are dangerous. Treads should be wide enough to give good footing. There should be no winding stair treads. Landings should be provided to break up any excessively long individual flight. Good railings make for safer use of stairs, and stairs of unusual width should have one or more center rails. During your inspection, check the condition of the slip resistant treads, and see that the handrails are securely mounted. Also check the landings at floor levels to be sure they are not being used for temporary storage.

Smokeproof Towers

A smokeproof tower is a stairway enclosure designed to prevent the movement of combustion products into it as occupants make their way out of the fire area into the tower. Access to smokeproof towers is gained by means of vestibules open to the outdoors through an exterior wall or from balconies overhanging an exterior wall. Your inspection should ensure that the egress path through the tower to the outdoors at street level is unobstructed throughout and that all guardrails on balconies or in vestibules are secure.

Ramps

Ramps, enclosed and otherwise arranged like stairways, are sometimes used instead of stairways where there are large crowds. They are required where a stair would have less than three steps. Ramps must have a very gradual slope to be considered safe exits. Again, see that they are unobstructed, and that their guardrails are secure.

Exit Passageways

Any hallway, corridor, passage, tunnel, or underfloor or overhead passageway may be considered an exit component, provided it complies with the requirements for exits. It must be of fire resistive construction, and openings into it are limited to those used by occupants for egress purposes. In addition to watching for the usual problems of means of egress, make sure that exit passageways are not being used for purposes inconsistent with the requirements for exits; for exam-

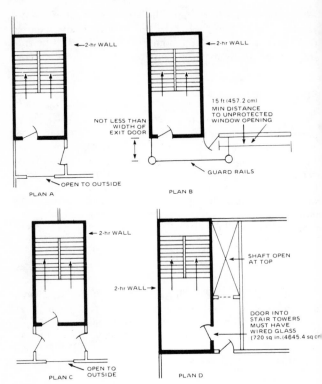

Figure 7-2. Four variations on smokeproof towers.

ple, vehicular traffic or pipeline transmission of hazardous materials.

Fire Escape Stairs

External fire escape stairs are no longer permitted in new buildings by the *Life Safety Code*, but there are still many older buildings that have them. Your inspection should ensure that fire escape stairs are unobstructed, in good repair, securely fastened to the building, and in operating condition.

Escalators and Moving Walkways

Escalators may be designated as exits if they are enclosed like exit stairs and meet the tread width and riser height requirements of exit stairs. Generally, they are not installed in a manner that would qualify them as exits.

Moving walkways may also be qualified as means of egress if they meet certain requirements — those for ramps if they are inclined, or those for passageways if level.

EXIT LIGHTING AND IDENTIFICATION

Lighting of the means of egress is necessary to ensure that occupants can quickly evacuate the building. This lighting must be continuous as long as the occupancy conditions require that the means of egress be available for use.

In some facilities, an independent source of power is available to be cut into the building's system automatically upon failure or open circuit of the primary source of power. Battery-operated electric lights and portable lights or lanterns may be used to provide emergency illumination of exits.

During routine inspections, you should observe the exit lighting system. Periodically test or witness the testing of emergency exit lighting equipment.

It is important that exits be adequately identified with exit signs and, in some cases, signs directing the occupant to an exit. In some cases it may be necessary to identify doors, stairs, or passageways that are not exits but that might be mistaken

---

Note: Elevators, ropes and ladders, and windows are not recognized as exits. However, windows meeting certain criteria may be used to provide access to fire escapes in existing buildings.

for exits. The *Life Safety Code* specifies sign size, letter dimensions, and illumination levels for exit signs.

EXIT DRILLS

The fire exit drill is a necessary exercise to familiarize occupants with the procedure to follow and the route to take in case of an emergency. In schools, they are absolutely essential. In some occupancies, such as hospitals and nursing homes, participation is often limited to employees. Evacuation plans should provide for locations outside the building and a safe distance away for occupants to assemble and be accounted for. In occupancies where activities are carried on 24 hours a day, the drills should be conducted on all shifts. Drills should be held unannounced and should simulate actual fire conditions. Your part in the exercise may be to assist by monitoring the drill, or you may be required to take charge of its planning and execution.

The information in this chapter is general, and there are many specific provisions, depending on the class of occupancy. You should become well informed as to the local requirements and the principal means of escape from fire as a total system. NFPA *101*, the *Life Safety Code*, covers means of egress in detail.

BIBLIOGRAPHY

McKinnon, G. P. ed., *Fire Protection Handbook*, 15th ed., National Fire Protection Association, Quincy, MA, 1981. Section 6, Chapters 1 and 2 assess life safety in buildings and discuss the concepts of egress design.

NFPA Codes, Standards, and Recommended Practices. (See the latest *NFPA Codes and Standards Catalog* for the availability of current editions of the following documents.)

NFPA 70, *National Electrical Code®*. Part of this code spells out the circuit requirements for emergency lighting systems.

NFPA 101, *Code for Safety to Life from Fire in Buildings and Structures*. In this code are specifications for construction and equipment used in recognized exits. Both basic requirements and modifications determined by occupancy are given.

Chapter 8

ELECTRICAL SYSTEMS

Although the examination and testing of electrical systems is best carried out by qualified electrical inspectors, the fact that fires can result from electrical system deficiencies, damage, or misuse makes it essential that fire inspectors be aware of the signs and symptoms of potential fire hazards from electrical systems.

In addition, fire inspections are carried out more frequently than electrical inspections by qualified electricians, making it more likely that potential faults can be detected before a failure occurs. In reporting electrical problems, always include a recommendation that the defect be rectified by a qualified person.

When fires of an electrical origin are reported, they may be placed in one of four broad categories — those caused by worn-out electrical equipment, by the improper use of equipment, by accidental means, and by defective installations. By learning to recognize visible signs of potential trouble, you will go a long way through your inspections toward eliminating electrical malfunctions as a cause of fire.

WIRING AND APPARATUS

Electrical fires are due principally to arcs and overheating. Arcing occurs when electrical current or energy attempts to take a short route, and can occur over small breaks in a conducting wire or from a conductor to metal in very close proximity. An arc produces sufficient heat to ignite nearby combustible material. Although conditions creating an arc usually will cause protective devices, such as fuses and circuit breakers, to operate, making heat exposure brief, intermittent arcing can sometimes occur without tripping such devices. Arcs may ignite combustible material such as insulation, fuse metal conductors, and produce sparks. Overheating is more subtle and is slower to cause ignition, but is equally capable of fire potential. Dangerous heat is generated in conductors and other electrical equipment when the current carried is in excess of rated capacity. Overloading deteriorates insulation and may ignite nearby combustible materials. Insulation deterioration by overheating can also lead to arcing of the conductors.

Common Faults

Many common faults are obvious; others are not.

Conduits, Raceways, and Cables: Among the obvious faults are badly deteriorated and improperly supported conduits, raceways, and cables. Where they enter boxes, cabinets, and other equipment, they should be terminated in proper fittings that hold them securely in place. Cables should be protected from mechanical damage where they pass through walls.

Not so immediately obvious are overloaded cables. They may be warm or even hot to the touch, depending upon the degree of overload. If cables feel abnormally warm, the reason should be investigated and corrective action taken.

Circuit Conductors: Like cables, branch circuit conductors need to be properly supported along their length and where they terminate in junction, switch, and outlet boxes. Conductors also should not be exposed to excessive external heat, which will hasten the deterioration of the insulation. Circuit conductors may also be subject to electrical overload when improperly protected. Overload protection will be discussed later in this chapter.

Flexible Cords: Several faulty practices involve flexible cords. They should never be used in place of fixed wiring, but sometimes they are. Flexible cords should not be nailed, tacked, or stapled to woodwork or tied or taped to pipes. Flexible cords should not be spliced, but should terminate in an approved connector or terminal. The electrical connection at the termination of a flexible cord should not be relied upon to provide mechanical support. Cords should be clamped or knotted where they enter sockets, fixtures, or appliances to prevent mechanical stress from being placed on the connection itself. Frayed cords should be replaced. Flexible cords should never be left where they can be damaged by vehicles or carts, or tucked under carpets.

Boxes and Cabinets: Outlet, switch, and junction boxes and cabinets are provided to protect the equipment and connections that they house. On your inspection tours, note that all such boxes are equipped with the proper cover. Boxes and cabinets are made with "knockouts," regular weak spots, which can be removed to permit entrance of a cable. No more knockouts are to be removed than are necessary to accommodate the conductors entering the box. The number of wires in a box or cabinet must not exceed the number for which it was designed. When observing outlet and switch boxes, look for cracked or broken switch and outlet assemblies. If any are found, they should be replaced promptly.

Switchboards and Panelboards: Other sources of electrical hazards are switchboards and panelboards, such as are found in theaters and in some industrial operations. Generally, they have exposed live parts from which occupants must be protected. There should be a cage or barrier around the live parts to prevent people from coming into contact with them. These boards usually employ bus bars, which should be adequately supported. Your inspection should include observations for deterioration, dirt, moisture, and poor maintenance.

Lamps and Light Fixtures: Light fixtures are subject to deterioration and poor maintenance. With age, the insulation on fixture wires may dry, crack, and fall away, leaving bare conductors. Sockets may become defective, and the fixtures themselves may become loose in the mountings. Fixtures should not be mounted directly on combustible ceilings.

Lamps often operate at temperatures sufficient to ignite

combustible material; therefore, they should be mounted with sufficient clearance so that continuous operation will not ignite paper, cloth, or other combustible material used as shades or that may be nearby. Unguarded portable lamps may ignite ordinary combustibles if placed in contact with them. A broken lamp may cause an ignition if the atmosphere is laden with combustible dust in suspension or with flammable liquid vapors.

Grounding

Hazardous voltages can be impressed on electrical distribution systems and equipment by lightning, accidental contact with a high voltage source, surface leakage due to conductive dirt or moisture, and breakdown of insulation on conductors. If the affected equipment is permitted to "float" at a dangerous voltage, then a person coming in contact with it and a point of lower potential, such as ground, will receive a serious, if not fatal, shock.

By grounding one conductor of the electrical circuit and then grounding all metals that may come in contact with a live conductor, a fault to ground occurs if the ungrounded circuit conductor should become accidentally grounded. The fault current finds a path through the grounded conductor and can cause the operation of an overcurrent device in the ungrounded circuit conductor, eliminating the dangerous condition.

Metal cable armor, raceways, boxes, and fittings as well as the frames and housings of electrical machinery must be grounded. Cord-and-plug-connected appliances, such as washers, dryers, air conditioners, pumps, etc., and certain electric tools must be grounded through a third contact in the line plug.

A metallic underground water piping system must be used as the grounding electrode where it is available and the buried portion of the pipe is more than 10 feet long. A metal underground water pipe grounding electrode, however, must be supplemented by an additional electrode as assurance of

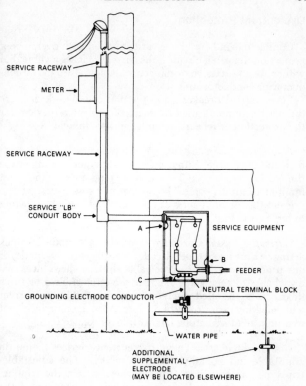

Figure 8-1. Grounding at a typical small service (AC, single-phase, three-wire, 115/230 v).

maintaining the integrity of the grounding electrode system.

Ground clamps and connectors should be checked periodically to ensure that they are tight and that the ground connection is being maintained. When new electrical machinery or equipment is installed on the premises, be sure that it too has been properly tied into the grounding system.

Overcurrent Protection

Conductors and equipment are provided with overcurrent protection to open a circuit if the current reaches a value that will cause an excessive or dangerous temperature in the conductor or conductor insulation.

Fuses, circuit breakers, and thermal overload devices are the most commonly used overcurrent protective devices for the protection of feeders, circuits, and equipment.

Plug Fuses: There are two types of plug fuses — the Edison base and the S-type. Either may be of the quick acting or the time delay type. The Edison base plug fuse is the most familiar to most people The S-type plug fuse is designed to prevent the use of pennies or other types of bridging schemes to bypass the fuse.

Cartridge Fuses: Cartridge fuses also are made in quick acting or time delay types. In addition, they are also made for one-time use or with renewable links. There are two drawbacks to renewable link cartridge fuses. Two or three links can be installed, thereby increasing the fusing current and defeating the purpose of the fuse. Also, upon replacement of a link, the fuse can be left with loose connections.

Circuit Breakers: These overcurrent devices come in adjustable-and nonadjustable-trip models. The adjustable-trip type can be set to trip anywhere within the tripping range of the device. The overload trip mechanism of the nonadjustable-trip type is thermally actuated; therefore, high ambient temperatures can reduce the current required to trip the breaker.

Thermal Overload Devices: These devices are not intended for protection against short circuits, but for protection against currents of a lower magnitude — for example, as thermal overload protection for an electric motor.

Current-limiting Overcurrent Devices: These devices are

generally used in high-current situations. When they inter-
rupt a specified current, they will consistently limit the short-
circuit current in the circuit to a specified magnitude substan-
tially less than that obtainable in the same circuit if the device
were replaced with a solid conductor having comparable impedance.

Ground-Fault Circuit-Interrupters: These devices sense
when a current — even a small current — passes to ground
through any path other than the proper conductor. When this
occurs, the GFCI trips almost instantly, stopping all current
flow in the circuit. GFCIs are extremely important for life
protection in wet locations.

INDUSTRIAL EQUIPMENT

Transformers

Large dry and fluid-filled transformers are used in certain
industrial applications, such as in industrial furnaces. Where
oil-filled transformers are installed on, or adjacent to, a
building or combustible material, there should be safeguards.
These include separation by fire resistive barriers or
enclosures to confine the oil if the transformer tank should
rupture. Examples are fire resistive dikes, curbed areas, or
basins or trenches filled with coarse crushed stone. Safeguards
also include water spray systems (see Chapter 28).

For the indoor installation of certain transformers, a standard vault is required. The vault should have walls and ceiling with a 3-hour fire resistive rating. Vaults require
drainage, a door sill raised enough to prevent the spread of
transformer oil, and a fire door. Openings for natural ventilation are required, but should not expose combustible
materials, windows, or exit stairs.

Motors

Sparks or arcs created when a motor winding short-circuits may ignite nearby combustibles. Bearings may overheat because of improper lubrication. Dust deposits or accumulations of textile fibers may prevent normal heat dissipation from the motor.

Your inspection regarding motors should indicate that there are no combustibles in the immediate vicinity of the motor or its controls, that the equipment is properly cleaned and maintained, and that it is provided with proper overcurrent protection. Motors are designed to operate normally without overheating. A hot casing can indicate a potential problem, and a closer examination should be made.

HAZARDOUS AREAS

Hazardous areas are those in which flammable liquids, gases, combustible dusts, or readily ignitible fibers or flyings are present in sufficient quantity to represent a fire or explosion hazard. In these areas, special electrical equipment is necessary. Complete definitions of the several classes and divisions of hazardous (classified) locations and methods of wiring and types of electrical equipment to be used in each are covered in the *National Electrical Code*.

Class I, Division 1

These are locations in which ignitible concentrations of flammable gases or vapors exist under normal conditions, or in which ignitible hazardous concentrations of such gases or vapors may exist frequently because of repair or maintenance operations or because of leakage, or in which breakdown or faulty operation of equipment or processes might also cause simultaneous failure of electrical equipment. Electrical equipment used in these locations must be of the explosion-proof type or purged and pressurized type approved for Class I locations.

Class I, Division 2

These are locations in which volatile flammable liquids or flammable gases are handled, processed or used, but in which the liquids, vapors, or gases will normally be confined within closed containers or systems from which they can escape only in case of accidental rupture or breakdown or abnormal operation of equipment; or in which ignitible concentrations of gases or vapors are normally prevented by positive mechanical ventilation and which might become hazardous through failure or abnormal operation of the ventilating equipment; or that are adjacent to but not cut off from Class I, Division 1 locations and to which ignitible concentrations of gases or vapors might occasionally be communicated.

Class II, Division 1

These are locations in which combustible dust is, or may be, in suspension in the air continuously, intermittently, or periodically under normal operating conditions in quantities sufficient to produce explosive or ignitible mixtures; or where mechanical failure or abnormal operation of equipment might cause explosive or ignitible mixtures to be produced and might also provide a source of ignition through simultaneous failure of electrical equipment; or in which combustible dusts of an electrically conductive nature may be present.

Class II, Division 2

These are locations in which combustible dust will not normally be in suspension in the air in quantities sufficient to produce explosive or ignitible mixtures and where dust accumulations are normally insufficient to interfere with normal operation of electrical equipment or other apparatus; and in which dust may be in suspension in the air as a result of infrequent malfunction of handling or processing equipment and dust accumulations resulting therefrom may be ignitible by abnormal operation or failure of electrical equipment or other apparatus.

Class III, Division 1

These are locations in which easily ignitible fibers or materials producing combustible flyings are handled, manufactured, or used.

Class III, Division 2

These are locations in which easily ignitible fibers are stored or handled, except in process of manufacture.

STATIC ELECTRICITY

Precautions against sparks from static electricity are needed in locations where flammable vapors, gases, or dusts are present or where there are easily ignited materials. The presence of static charges may be detected by a simple neon tube tester of the type commonly used for testing automobile spark plugs. The neon tube glows when one lead is brought into contact with a body charged with static. Other instruments for the purpose are the electrostatic voltmeter, gold leaf electroscope, or vacuum tube detector.

Where dangerous static conditions exist, measures appropriate to bring the hazard under reasonable control are humidification, bonding, grounding, ionization, or a combination of these methods.

Humidification

Humidity alone is not a completely reliable means of eliminating static charges, but to reduce the danger of static, the relative humidity should be high. Some industrial operations cannot be performed at humidities high enough to obviate static danger. A practical rule is to maintain relative humidities as high as possible without undue hardship, even up to 75 percent.

Bonding and Grounding

Bonding is done to minimize differences in electrical potential between metallic objects. There is practically no potential difference between two metal objects connected by a bond wire. The current a bond wire has to carry is generally small.

The term grounding is used to describe connections made to minimize differences in electrical potential between objects and the ground. You must recognize that, in such cases, the ground wire may carry current from power circuits that is much larger than static current.

Ground connections should be tested when the installation is made and frequently thereafter to detect impairment by corrosion, loose connections, or injury.

Flowing gases, liquids, or granular solids, like sand, generate static. When gasoline, for example, is transferred from a drum to a can, it is important that the drum and can be effectively bonded together by a tube that is electrically conductive and firmly in contact at both ends, or that a grounding wire be securely attached to both containers before transfer of liquid is started.

In general, the bonding and grounding connections should be of substantial construction, so that they are not easily broken, and installed so that the inspector can readily see that they are in place and intact.

Ionization

Ionization is the process of increasing the conductivity of

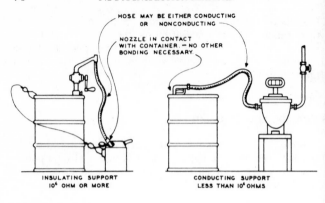

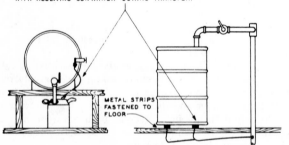

Figure 8-2. Recommended methods of bonding flammable liquid containers during container filling.

air, so that static charges will be conducted away.

One technique employs the tendency of static to concentrate on the surface of least radius of curvature, such as a sharp point. A metal bar with needle points (static comb) or with metallic tinsel provides a method of removing static from moving sheet materials.

Another technique is use of a so-called electrical neutralizer. This produces an alternating electrical field through which the electrified sheet material has to pass.

A flame ionizes surrounding air, providing a technique used on printing presses.

Ionization is also produced by alpha radiation from a radioactive surface.

The hazards introduced by these various techniques must be considered as well as their respective effectiveness in removing static charges.

LIGHTNING PROTECTION

A lightning protection system embodies the following elements:

• Air terminals to pass the discharge, high enough above the structure to avoid fire from the arc.

• Conductors following a direct path to ground without sharp bends or loops, providing a rudimentary cage of low impedance around the structure.

• Ground connections of low resistance distributed symmetrically.

• Bonding to conductors of sizable and nearby metal objects to avoid side flashes.

• Independent grounding of other sizable metal objects.

• Strong mechanical construction and high resistance to corrosion in the system, which must be expected to be in working condition for long periods with little attention.

Grounding arrangements will depend on the soil. In moist clay and similar soil of relatively high conductivity, ground rods extended not less than 10 feet vertically into the soil may serve. In soil that is largely sand, gravel, or stones, more extensive ground electrodes may be required, consisting of driven rods or pipes. Such pipes should be marked for the convenience of the inspector to show the depth of the pipe or rod. Another grounding method is to lay lengths of conductors in trenches radially from the structure. Such ground trenches usually should be 12 feet long and need not be more than

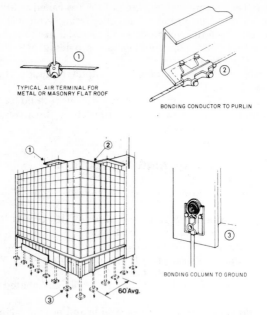

TYPICAL AIR TERMINAL FOR
METAL OR MASONRY FLAT ROOF

BONDING CONDUCTOR TO PURLIN

60′ Avg.

BONDING COLUMN TO GROUND

Figure 8-3. Grounding and bonding of lightning down conductors. Water pipe grounds (if pipes are metallic) can be made at 1, 2, or 3.

3 feet deep. Where soil is very dry or where bedrock is near the surface so that digging more than a foot is impractical, a conductor is buried to encircle the structure and to bond together all of the down connections. A conductor is usually brought down where a water pipe enters the building and is grounded to the pipe.

SI UNITS

The following conversion factor is given as a convenience in converting to SI units the English unit used in this chapter.

1 ft = 0.305 m

BIBLIOGRAPHY

Electrical Appliance and Utilization Equipment Directory, Underwriters Laboratories Inc., Northbrook, IL, issued annually. This is a listing of the electrical appliances and devices that have been tested and found safe for use.

Electrical Construction Materials Directory, Underwriters Laboratories Inc., Northbrook, IL, issued annually. A directory of tested and listed construction materials such as circuit breakers, wires, transformers, industrial control equipment, electrical service equipment, and fixtures and fittings.

Hazardous Location Equipment Directory, Underwriters Laboratories Inc., Northbrook, IL, issued annually. A listing of electrical components and equipment that have been tested and listed for use in hazardous atmospheres.

McKinnon, G. P. ed., *Fire Protection Handbook*, 15th ed., National Fire Protection Association, Quincy, MA, 1981. Section 7, Chapter 2, discusses electrical systems and appliances in detail.

NFPA Codes, Standards, and Recommended Practices. (See the latest *NFPA Codes and Standards Catalog* for the availability of current editions of the following documents.)

NFPA 70, *National Electrical Code*. Contains the requirements for safe electrical systems.

NFPA 77, *Recommended Practice on Static Electricity*. Discusses the nature and origin of static electricity and methods of its mitigation and dissipation.

NFPA 78, *Lightning Protection Code*. Contains the requirements for protection against damage caused by static electricity.

NFPA 325M, *Fire Hazard Properties of Flammable Liquids, Gases and Volatile Solids*. A tabulation of some 1,500 substances listed alphabetically by chemical name.

Chapter 9

HEATING SYSTEMS

Heat producing appliances and systems operate at temperatures above the ignition temperatures of many common combustible materials. Therefore, see that there are adequate clearances between combustibles and the heating appliance, its ducts, and its pipes. Malfunctions in these devices can result in their becoming sources of ignition of hostile fires. This chapter deals with systems used for heating the interior of buildings. Systems used for industrial processes and power generation are discussed in Chapter 22.

It will be helpful to you in conducting your inspections if you are familiar with the operation and safety features of the various equipment currently in use.

FUELS AND FIRING METHODS

The most widely used fuels for heating systems are coal, oil, and gas.

Coal

Coal used for heating fuel covers a wide range from anthracite — a hard, dense, clean burning coal — through subbituminous coal, which is subject to spontaneous ignition under certain storage conditions, to coke, which is a product of the destructive distillation of coal.

Coal is fired either manually or automatically by stokers or pulverized coal burners.

The only automatic feature about a hand-fired or manually fired furnace is that the draft is regulated automatically. It can be controlled by a room thermostat, by steam pressure, or by water temperature. However, the fire is continuous, and overtemperature conditions are difficult to control.

Mechanical stokers are capable of handling from 10 to 1,200 pounds per hour. Basically, there are four types of mechanical stokers — underfed, overfed, traveling and chain grate, and spreader with fired grate or continuous ash discharge.

Pulverized coal systems are used primarily in boiler-furnace applications (see Chapter 21).

Oil

There are a number of grades of fuel oil in use today. No. 1 and No. 2 fuel oils as well as kerosene, range oil, furnace oil, star oil, and diesel oil are known as distillates. Nos. 4, 5, and 6 fuel oil and Bunker C oil are called residuals. When inspecting properties having oil-fired heat producing equipment, look for any evidence of leaks.

There are two methods of firing oil — vaporization and atomization. While the gun-type atomizing burner is most common today, many of the earlier vaporization type of burners are still in service.

Vaporization burners are of several designs. The sleeve type and natural draft pot burners are used in residential heating and in cooking stoves. Forced draft pot burners may be used in central heating furnaces. Vertical rotary wall flame vaporizing burners are used in residential boilers and in furnaces for central heating.

Likewise, there is a variety of atomizing burners. High pressure gun burners are generally designed to burn No. 2 fuel oil, though some are made to burn No. 4. The low pressure gun burner differs from its high pressure counterpart. It includes an air pump to supply compressed air for atomization, and the oil and air are delivered to the nozzle at a pressure of 15 psi or less. The horizontal rotary cup oil burner atomizes the oil by spinning it in a thin film from a horizontal rotating cup and injecting high velocity primary air into the oil film through an annular nozzle that surrounds the rim of the cup. Secondary air for combustion is supplied by a separate fan, which forces air through the burner wind box.

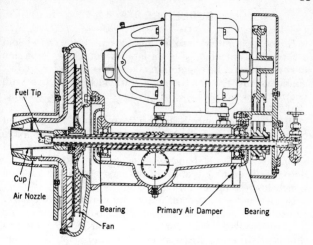

Figure 9-1. Horizontal rotary cup oil burner. (National Fuel Oil Institute)

Gas

LP-Gas, an LP-Gas-air mixture, and natural gas are also used to fuel heat-producing devices. Because LP-Gas vapors are heavier than air, your inspection of such equipment in below grade indoor locations should be especially critical. Practically all residential and commercial gas appliances use injection burners in which air is mixed with gas before it reaches the burner ports or point of ignition.

In luminous or yellow flame burners, air is supplied at the point of combustion; it is not premixed with the gas.

In power burners, either gas or air or both are supplied at pressures higher than the line pressure of the gas and the atmospheric pressure of the air.

Catalytic burners permit combustion of the gas at temperatures well below the normal ignition temperature of the gas-air mixture.

BURNER CONTROLS

If fuel is permitted to collect in the combustion chamber of a furnace in the absence of an ignition source, an explosion may result if a delayed ignition takes place. Therefore, safety considerations require that fuel burners be equipped with controls to cut off the fuel supply in the event of a malfunction.

Primary safety controls shut off the fuel supply in the event of flame or ignition failure. Interlock circuits are provided to shut off the fuel supply if an induced or forced draft fails, if atomization fails, if dangerous fluctuations in fuel pressure occur, or if oil temperature in burners requiring heated oil falls below the minimum required.

Burners not equipped for automatic restarting after safety controls have operated to extinguish the burner flame should be provided with means for manual restarting. Remote fuel shutoffs, either electrically or manually controlled, are generally recommended. They should be located so that it is not necessary to pass the burner in order to operate them.

CENTRAL HEATING APPLIANCES

Boilers either heat water or generate steam, which is distributed through tubing, pipes and radiators to heat the surrounding air. Steam boilers that operate at pressures not exceeding 15 psi are known as low pressure boilers. Hot water boilers operate at pressures up to 160 psi and temperatures up to 250°F. Safety devices for boilers include means for pressure relief in the event boiler pressure attempts to exceed the maximum operating pressure, low water shutoff, and high temperature and high pressure shutoffs. Fire problems of boilers generally involve clearances from combustibles and placement on the floor (noncombustible or protected combustible floors).

Central warm air furnaces are either the gravity type or the forced air type. Gravity furnaces are mounted on the floor and heat only the spaces above them. Ductless gravity furnaces are called pipeless furnaces. Floor furnaces are a form of gravity furnace that is suspended from the floor and heats the area above. Gravity furnaces should be equipped with high temperature limit controls that shut off the fuel supply when the temperature of the discharge air reaches a predetermined level.

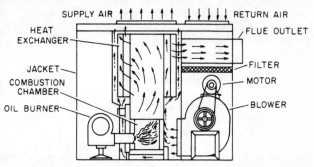

SUPPLY AIR RETURN AIR
HEAT EXCHANGER
FLUE OUTLET
JACKET
FILTER
COMBUSTION CHAMBER
MOTOR
OIL BURNER
BLOWER

Figure 9-2. An oil-fired forced air furnace. (National Fuel Oil Institute)

Forced warm air furnaces are equipped with plenums, which may become hot enough to ignite adjacent combustibles. Such furnaces should be equipped with a limit control to shut down the fuel supply when the temperature in the plenum or at the entrance to the supply duct reaches a predetermined level. With all warm air furnaces, it is important to maintain adequate clearances from combustibles.

Wall furnaces are self-contained indirect-fired gas or oil heaters installed in or on a wall. They supply heated air by gravity or with the aid of a fan, and they are either directly vented or vent- or chimney-connected. Wall furnaces should have high temperature limit controls.

Furnaces installed inside ducts are known as duct furnaces. They depend on air circulation from a blower that is not furnished as part of the furnace. They may be oil- or gas-fired or heated electrically and are equipped with a high temperature limit control.

A heat pump is a type of forced air heating system in which refrigeration equipment is used in a manner that heat is taken from a heat source and given up to the conditioned space when heat is wanted and is removed from the space when cooling is desired. Units employing supplemental heating units are equipped with an interlock that prevents the pump from operating when the indoor air circulating fan is not operating. Heat pumps are also equipped with temperature limit controls. The hazards of such an arrangement are those of electrical, refrigeration, and heating equipment.

UNIT HEATERS

Unit heaters are self-contained, automatically controlled, chimney- or vent-connected air heating appliances, equipped with a fan for circulating air. They may be floor mounted or suspended, and they are equipped with limit controls. Unit heaters that are connected to a duct system may be considered as central heating furnaces and should be provided with the same safeguards.

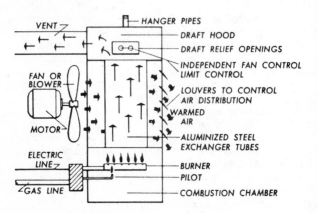

Figure 9-3. A typical gas-fired unit heater.

HEAT DISTRIBUTION

Warm Air Systems

Horizontal supply ducts, vertical ducts, risers, boots, and register boxes can reach hazardous temperatures. It is essential, therefore, that safe clearances to combustibles be maintained. NFPA 90B, Installation of Warm Air Heating and Air Conditioning Systems, contains required clearances for many warm air ductwork configurations. One register or grill must be installed without a shutter and without a damper in the duct leading to it. This is to prevent excessive heat from developing in the duct system. Automatic oil- or gas-fired systems having suitable temperature limit controls need not meet this requirement. Neither must systems in which shutters and dampers cannot shut off more than 80 percent of the duct area.

The best way to conduct return air to the furnaces is through continuous ducts. However, each vertical stack may be connected to registers on one floor only. If vertical stacks were permitted to connect to registers on more than one floor, they would become paths for the ready communication of heat, smoke, and toxic products of combustion from one floor to another.

Steam and Hot Water Systems

Hot water pipes and radiators in systems operating with a maximum temperature of 150°F require no installation clearances. Pipes and radiators supplied with hot water up to 250°F or steam at a pressure up to 15 psig require a clearance of 1 inch from combustibles. Where these pipes pass through a floor, wall, or ceiling, the clearance at the opening through the finish boards may not be less than ½ inch and must be covered with a plate of noncombustible material.

INSTALLATION

A major consideration in the installation of any heat producing appliance is its effect on nearby combustibles. It is possible for wood and other combustibles to ignite at temperatures well below their usual ignition temperatures if they are continually exposed to moderate heat over long periods of time. For this reason, installation clearances and steps to insulate combustible surfaces are of the utmost importance. Extensive information on clearances is given in Section 7, Chapter 3 of the *NFPA Fire Protection Handbook*.

Listings of tested heating equipment indicate the suitable material upon which the equipment may be mounted. Examples are combustible floors, fire resistive floors extending specific distances beyond the equipment, masonry floors, or metal over wood. These are covered rather thoroughly by type of appliance in the the *Fire Protection Handbook* and by the specific manufacturer's model in laboratory listings.

Another installation requirement is that ample air be provided for combustion and ventilation. An oxygen-starved fire is a dirty and inefficient fire, which will lead to maintenance problems rather quickly. Additional air for ventilation helps to carry away the heat that develops on the surface of the equipment. In relatively large, open areas, the air supply poses no particular problem. However, in furnace or boiler rooms made of fire resistive construction and equipped with fire doors, it may be necessary to make provisions for supplying the necessary air. NFPA 54, *National Fuel Gas Code*, contains specific recommendations on how to supply the air required for combustion and ventilation.

CHIMNEY AND VENT CONNECTORS

Chimney and vent connectors are those lengths of pipe or other type of conduit that connect the heat producing appliance to the chimney or vent. Connectors are made of noncombustible corrosion resistant material, such as steel or refractory masonry. They must be able to withstand flue gas temperatures and resist physical damage. Connectors must be short, well fitted and supported, continuously pitched toward the chimney or vent, and have adequate clearance from combustibles.

VENTS

Vents are used with specific types of heat producing equipment. Type B gas vents are used with such devices as gas-fired furnaces, ranges, and water heaters. Type BW vents are listed for use with vented wall furnaces, and single wall metal pipe is suitable for venting residential-type, low heat gas appliances and outdoor incinerators.

In buildings requiring protection of vertical openings, vents are enclosed in fire resistive construction. For buildings less than four stories high, the construction must have a 1-hour rating. For buildings four stories or more, the construction must have a 2-hour rating.

CHIMNEYS

Chimneys are classified into three major types — masonry, factory-built, and metal.

Masonry Chimneys

A masonry chimney should be inspected for its entire length so far as it may be accessible. The inside of the chimney may be examined by placing a mirror in a connector opening and catching the sunlight.

On the roof, note the condition of the mortar, chimney lining, and flashing, and look for evidence of cracking or settling. Note the number of flues. In the attic, check for cracks and loose mortar; do the same in the basement. On other floors, check chimney connections.

One simple method of determining that a chimney is unsafe

is to hold your hand against it while it is in use. If it is too hot for comfort, no combustible material should be permitted to come in contact with it.

If mortar has begun to fall out from between the bricks, openings may be expected to develop all the way through the wall. If a sharp instrument can be pushed through the wall, it is time to rebuild the chimney. The tops of chimneys are the most likely places to need rebuilding.

Factory-Built Chimneys

Factory-built chimneys are lightweight assemblies and good draft producers. Some types resemble Type B gas vents, but are larger and heavier. The materials used in their construction meet certain requirements for heat and corrosion resistance and physical tests, where applicable, for crushing and for freezing and thawing.

Factory-built chimneys are available in listed assemblies for low and medium heat appliance service.

Metal Chimneys

Metal chimneys are suitable for all classes of appliances, but are not subjected to safety testing of any kind. The major hazard of these chimneys is inadequate clearance to combustibles where they penetrate ceilings and roofs. They may be located outside of buildings, but not inside one- or two-family dwellings or buildings of wood frame construction. The conditions under which metal chimneys can be used are quite limited and are spelled out in detail in NFPA 211, *Chimneys, Fireplaces, Vents, and Solid Fuel Burning Appliances.*

A CHECK LIST OF CHIMNEY DEFECTS

A list of common and important defects which, individual-

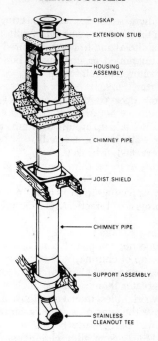

- DISKAP
- EXTENSION STUB
- HOUSING ASSEMBLY
- CHIMNEY PIPE
- JOIST SHIELD
- CHIMNEY PIPE
- SUPPORT ASSEMBLY
- STAINLESS CLEANOUT TEE

Figure 9-4. A typical factory-built chimney. (Metalbestos Div., Wallace-Murray Corp.)

ly or in combination, constitute sufficient reason for requiring repairs to or complete rebuilding of a masonry chimney.

1. General structural unsoundness of design or proportionate dimensions of the chimney.

2. Evidence of settling or cracking due to inadequate footings or other causes.

3. Chimney rests upon or is wholly or partly carried by wooden floors, beams or brackets or is hung by metal stirrups from wooden construction. Chimney used to support any wooden floor or roof beams.

4. Chimney increases in size, has projecting masonry, or is set back within 6 inches above or below rafters or roof joists.

5. Chimney unlined and less than required wall thickness.

6. Masonry unbonded or improperly bonded or sections not properly anchored or reinforced.

7. Weak mortar.

8. Old mortar decayed due to action of flue gases, or weathering. Chimney not properly finished at top.

9. Brickwork not laid up around the lining (lining dropped into place after walls are constructed).

10. Linings cracked or broken.

11. No fire clay or metal thimbles provided at openings for connectors.

12. Connector openings in more than one story for a single flue without provision for effectively closing unused openings.

13. Flues show leakage in smoke test.

14. Flue linings not complete from 8 inches below connector openings to top of chimney.

15. Reduction in cross-sectional area of a flue at any point.

16. Flue at greater than 30-degree angle with vertical.

17. Chimney height less than 3 feet above a flat roof or less than 2 feet above the edge of a gable or hipped roof.

18. Woodwork, particularly beams and joists, within 2 inches of outside surface of all of chimney.

19. Combustible material or construction near ash pit or cleanout doors; access to such doors blocked.

SI UNITS

$$1 \text{ in.} = 25.4 \text{ mm}$$
$$1 \text{ psi} = 6.895 \text{ kPa}$$
$$\%(\degree F - 32) = \degree C$$

BIBLIOGRAPHY

ASME Boiler and Pressure Vessel Code, Section IV, American Society of Mechanical Engineers, New York, 1974. Defines low pressure boilers and hot water boilers.

Chimney for Medium Heat Appliances, ANSI/UL 959, Underwriters Laboratories Inc., Northbrook, IL, 1976. A safety test standard for chimneys.

Gas and Oil Equipment Directory, Underwriters Laboratories Inc., Northbrook, IL, issued annually. A directory of tested and listed oil- and gas-fired heating equipment.

Standard for Factory-Built Chimneys, ANSI/UL 103, Underwriters Laboratories Inc., Northbrook, IL, 1978. A testing standard for factory-built chimneys.

McKinnon, G. P. ed., *Fire Protection Handbook*, 15th ed., National Fire Protection Association, Quincy, MA, 1981. Chapter 3 of Section 7 is a detailed treatment of heating systems and appliances.

NFPA Codes, Standards, and Recommended Practices. (See the latest *NFPA Codes and Standards Catalog* for the availability of current editions of the following documents.)

NFPA 30, *Flammable and Combustible Liquids Code*. Requirements for tank storage, valves and fittings, container storage, industrial plants, bulk plants, and processing plants.

NFPA 31, *Standard for the Installation of Oil Burning Equipment*. Requirements for stationary and portable equipment, tanks, piping, and accessories.

NFPA 54, *National Fuel Gas Code*. Criteria for the installation, operation, and maintenance of gas equipment.

NFPA 70, *National Electrical Code*. Contains wiring requirements for furnaces.

NFPA 85, *Standard for Prevention of Furnace Explosions in Fuel Oil- and Natural Gas-Fired Watertube Boiler Furnaces with One Burner*. Design, installation, operation, and maintenance requirements.

NFPA 85B, *Standard for Prevention of Furnace Explosions in Natural Gas-Fired Multiple Burner Boiler-Furnaces*. Applies to the design, installation, and operation of these boiler-furnaces.

NFPA 85D, *Standard for the Prevention of Furnace Explosions in Fuel Oil-Fired Multiple Burner Boiler-Furnaces*. Outlines equipment requirements, sequencing of operations, and interlock and alarm requirements.

NFPA 89M, *Manual on Clearances for Heat Producing Appliances*. A compilation of recommended clearances.

NFPA 97M, *Standard Glossary of Terms Relating to Chimneys, Vents and Heat Producing Appliances*. Definitions of terms.

NFPA 211, *Standard for Chimneys, Fireplaces and Vents*. A standard on safe installation and use in residential, commercial, and industrial applications.

Chapter 10

AIR CONDITIONING AND VENTILATING SYSTEMS

Conditioned air is temperature and humidity controlled, cleaned and distributed to meet the requirements of the conditioned space. There are several types of air conditioning systems — those which provide washed or filtered, cooled dehumidified air in the summer and heated humidified air in the winter; those which provide only filtered, cooled dehumidified air; and those which provide only filtered, heated humidified air.

Air conditioning systems are divided into three segments: the air intake system, the conditioning equipment, and the distribution system.

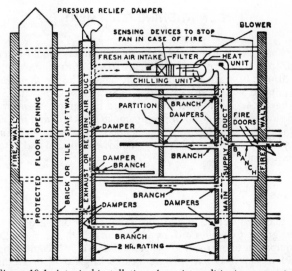

Figure 10-1. A typical installation of an air conditioning system in a building.

AIR INTAKE SYSTEM

Some systems mix fresh air with recirculated air, while others use fresh air exclusively. In either case, provision needs to be made for introducing fresh air to the system. Where there is potential for smoke, heat, or flames from an exposure fire outside the building to enter the air intake, it should be equipped with a fire door or damper that is operated automatically by fire and smoke detection equipment. Be sure that the door or damper is in working order and that the intake is free of accumulations of rubbish and debris that might constitute a hazard as well as restrict the flow of make-up air.

The opening of the air intake duct should be protected with a grill or screen that will prevent foreign materials from entering the system. Make sure that it is not broken, clogged, or missing.

CONDITIONING EQUIPMENT

Fans, heating and cooling units, and filters should be cut off from the rest of the building by room construction having a fire resistance rating of 1 hour. The enclosure should be equipped with a suitable door, which is kept closed. (See Chapter 4.) Equipment rooms should be kept clean and free of storage. Some form of fire or smoke detection and fire extinguishing system are recommended. Where such systems and devices are provided, they should be inspected and tested periodically.

Fans

Lack of lubrication and accumulations of dust are two of the greatest enemies of fans and motors. Both can cause the equipment to overheat to the point that it may become an ignition source. Often fans are located in places that are difficult to reach. Nevertheless, they should be accessible and included in your inspection program.

Heating and Cooling Equipment

The hazards of cooling equipment are related to the hazards of electrical installations and of the refrigerant itself. Proper wiring and grounding are discussed in Chapter 8. Most common refrigerants are at least slightly toxic; therefore, a leak in the system will be a hazard to health and life safety. Some refrigerants present a combustibility hazard. The greatest problem associated with refrigeration units is the explosion hazard of the pressurized refrigerant. Recommendations for the installation of mechanical refrigeration equipment are contained in the Safety Code for Mechanical Refrigeration issued by the American Society of Heating, Refrigerating, and Air Conditioning Engineers. The fire experience with air conditioning refrigerating units is generally good where the cooling equipment is properly installed and maintained. Your inspection should include checks of the quality of maintenance performed on the equipment and of the housekeeping in the vicinity of the equipment.

The hazards of air heating equipment depend on the method of heating used. Heating equipment is discussed in Chapter 9.

Air Cleaning Equipment

Basically, two types of air cleaning equipment are used in air conditioning systems — filters and electronic air cleaners. Both renewable, those that are washable and reusable, and disposable filters are used. The purpose of filters and air cleaners is to remove entrained dust and other particulate matter from the air stream. The filtered particles accumulate in the filter material or on the air cleaner collector plates, and, if ignited, may burn and produce large volumes of smoke. The products of combustion can be circulated throughout the building by the air distribution system, posing a threat to life safety.

There are two classes of filter. When clean, Class 1 filters do not contribute fuel when attacked by flame and give off only moderate amounts of smoke. Clean Class 2 filters will burn moderately when attacked by flame and emit moderate amounts of smoke.

Filters are either dry or the viscous impingement type, in which the filtering medium is coated with an adhesive to trap particulate matter. When the fibrous medium in a unit filter has trapped sufficient foreign matter to cause a certain predetermined pressure drop in the system, it should be discarded or washed with steam or a hot detergent solution. After being washed, the filtering medium in a viscous impingement type of filter should be recoated with an adhesive having a flash point no less than 325°F as measured in the Pensky-Martens closed tester.

Electronic air cleaners utilize electrostatic precipitation to remove particulate matter. Entrained particles pass through electrostatic fields and are collected either on a filter or on charged plates. Lethal voltages and currents are used in electronic air cleaners; therefore, they are equipped with interlocks that shut down the unit if a door or access panel is opened. As you make your inspection, note that the interlocks are intact and have not been bypassed.

DISTRIBUTION EQUIPMENT

Distribution of conditioned air throughout the building is through the duct system. During fire conditions, the same duct system can disperse toxic products of combustion throughout the building instead of breathable air.

Generally, ducts are fabricated of metal, masonry, or other noncombustible material. The American Society of Heating, Refrigerating, and Air Conditioning Engineers has published information on the installation of fibrous glass ducts. Underwriters Laboratories has a system for classifying duct materials according to flame spread, smoke development, and flame penetration. Class O materials have a flame spread and smoke development rating of O. Class 1 materials have a flame spread rating of 25 or less, with no evidence of continued progressive combustion, and a smoke development rating of not more than 50. Class 2 materials have a flame spread rating greater than 25 but not over 50, with no evidence of continued progressive combustion, and a smoke development rating of not more than 50 for the inside surface of the duct and not over 100 for the outside surface. Class O and Class 1 materials must pass a 30-minute flame penetration test, and Class 2 materials must pass a 15-minute flame penetration test.

Ducts can create both vertical and horizontal openings in structural fire barriers. Your interest, as a fire inspector, is in seeing that, where the ducts pass through fire barriers or fire walls, adequate firestopping has been provided to seal the space between the duct walls and the edges of the opening. If properly installed and firestopped, sheet metal ducts in the gages commonly used may protect an opening in a fire barrier for up to 1 hour. If a wall, partition, ceiling, or floor is required to have a fire resistance rating of more than 1 hour, then a fire damper is required in the duct to protect the opening. If the barrier is a fire wall, an automatic closing fire door suitable for a Class A opening (see NFPA 80, Standard for Fire Doors and Windows) is required.

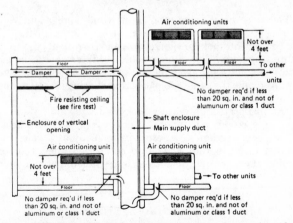

Figure 10-2. Sample fire damper requirements and examples of locations where dampers are not required.

SMOKE CONTROL

There are two recognized approaches to smoke control in buildings that make use of the air conditioning system. One requires that fans be shut down and smoke dampers in ductwork be closed during a fire. This is known as the passive approach. In the second approach — the active approach — the building's high velocity air conditioning system is used to prevent smoke migration from the fire area and to exhaust the products of combustion to the outdoors. When the active approach is used, smoke and fire dampers can be omitted from the system.

When smoke dampers are used in the passive form of smoke control, they are not controlled by heat sensitive devices, such as fusible links, but are controlled by smoke detectors. Motor-driven air control dampers may be used for smoke control. In this case, the smoke detector triggers an override of all other control functions, and the motor closes the damper.

MAINTENANCE

The keys to safe operation of air conditioning and ventilating systems are maintenance and cleaning. As you inspect the equipment, look for signs of rust and corrosion, especially on moving parts. Check the condition of filters, and examine return air ducts for accumulations of combustible dust and lint. Also note the condition of electrical wiring as well as any damage to ducts.

The fire protection devices associated with the system — fire suppression equipment, smoke control equipment, alarms, etc. — should be tested periodically as part of the maintenance program. If you do not witness these tests or conduct them yourself, records of the tests should be available to you.

VENTILATING SYSTEMS

Often, conditions warrant special ventilation systems; for

example, for the removal of flammable vapors, corrosive vapors or fumes, grease-laden air from cooking equipment, and combustible dusts. The hazards of such systems lie in the possibility of ignition of flammable materials or vapors from such sources as sparks generated by fans or foreign materials in the air stream or by overheated bearings.

To reduce the hazard of fire, fans should be of noncombustible construction, have provision for shutdown by remote control in case of fire, be accessible for maintenance, and be structurally sound enough to resist wear. When the exhaust material is flammable solids or vapors, fan blade and housing construction of nonferrous metal will minimize the possibility of spark generation.

These special exhaust systems should be independent of other ventilating systems and of one another. They should be vented directly to the outdoors by the shortest route and should not pass through fire walls or fire partitions.

SI UNITS

The following conversion factor is given as a convenience in converting to an SI unit the English unit used in this chapter.

$$\tfrac{5}{9}(°F - 32) = °C$$

BIBLIOGRAPHY

McKinnon, G.P. ed., *Fire Protection Handbook*, 15th ed., National Fire Protection Association, Quincy, MA, 1981. Chapters 4 and 5 of Section 7 are detailed treatments of air conditioning and ventilating systems and air moving equipment.

McKinnon, G.P. ed., *Industrial Fire Hazards Handbook*, 1st ed., National Fire Protection Association, Quincy, MA, 1979. Chapter 43 deals with the design, hazards, and equipment of air moving systems.

Safety Code for Mechanical Refrigeration, No. 15-78, 1978 ed., American Society of Heating, Refrigerating, and Air Conditioning Engineers, New York. This code describes the requirements for the installation of air conditioning equipment and ducts.

NFPA Codes, Standards, and Recommended Practices. (See the latest *NFPA Codes and Standards Catalog* for the availability of current editions of the following documents.)

NFPA 70, *National Electrical Code*. Contains wiring and grounding requirements for electrical equipment.

NFPA 80, *Standard for Fire Doors and Windows*. A standard on the use, installation, and maintenance of fire doors.

NFPA 90A, *Standard for the Installation of Air Conditioning and Ventilating Systems*. A standard to restrict the spread of smoke, heat, and fire through duct systems and minimize ignition sources.

Chapter 11

WASTE HANDLING SYSTEMS

Combustible waste is a fire waiting to happen. Although the waste generated within the course of a normal day's activity is rarely in itself a source of ignition, the very fact that waste can accumulate presents the potential for a growing source of fuel for fire starting elsewhere. Collecting and disposing of waste in an efficient and safe manner eliminates that potential.

Methods of handling waste are varied. They range from manual collection from waste baskets to sophisticated pneumatic systems servicing an entire building. Disposal of waste has the same wide ranges: from small land fills to huge multi-stage incinerators.

WASTE CHUTES AND HANDLING SYSTEMS

There are four types of waste chute systems: general access gravity type, limited access gravity type, pneumatic systems, and gravity-pneumatic systems.

General Access Gravity Type

This type of chute is an enclosed vertical opening passageway in a building leading to a storage or compacting room. Everyone in the building can use the chute.

Limited Access Gravity Chute

These are similar to the general access type except entry is gained by a key to a locked chute door.

Pneumatic Waste Handling System

This system depends on airflow to move waste to a collection point. It is mostly found in hospitals and health care buildings for carrying soiled linen, although it can be used for other wastes. In pneumatic systems fire dampers are located at any point where chutes penetrate fire-rated floor and ceiling assemblies, fire walls, etc. Find out where their dampers are located and make sure they are in good operating condition.

Gravity-Pneumatic Waste Handling System

This system uses a conventional gravity system to feed a collecting chamber which in turn feeds a pneumatic waste handling system.

Waste Chutes

Look at chutes as fire-prone channels capable of carrying flames and smoke throughout a building; consequently, they must be well constructed. Guidance on whether or not a chute is a good installation will be found in NFPA 82, *Incinerators, Waste and Linen Handling Systems and Equipment*.

Waste chutes are often considered as storage areas, since they can become clogged during use or clogged because they were not designed to handle the type of waste currently being fed into it. A clogged chute presents a smoke and fire hazard. Inquire if the chutes on the premises have a recent history of being jammed frequently with refuse. If that is the case, it may mean the chute is too small for the type of waste it handles, and a recommendation could be in order for a larger chute. Often a thorough study of the volume and type of waste and the habits of users of the chute is warranted.

Gravity-type chutes are constructed of either masonry, refractory-lined steel, or stainless, galvanized or aluminized steel. Floor openings through which a chute passes require fire-rated protection. Metal chutes need fire-rated enclosures for protection of the chute and floor openings. Make sure the enclosing envelope has not been damaged so that it is no longer a tight fire barrier.

Chutes should have sprinklers, and if you find chutes without them a recommendation is in order. Where sprinklers are installed, a good practice is to locate them where they can be inspected and maintained and yet be out of the reach of vandals and beyond the range of falling objects.

Chute Terminal Enclosures

The chute room or bin where the waste from chutes is collected must have a fire resistance rating equivalent to that required for the chute. Openings into the enclosure must be protected by fire doors suitable for Class B openings. Sprinklers in terminal rooms are a must. The seriousness of the fire hazard requires fast extinguishment capabilities.

Waste chutes should not dump directly into incinerators since this would make it possible for the chute to serve as a chimney as well. These have been largely discontinued or abandoned.

INCINERATORS

Incineration provides for the total destruction of wastes as rapidly as they are generated. It not only eliminates odors and unsightly vermin-attracting waste piles, it, more importantly, reduces potential fire hazard.

Incinerators are generally referred to as being of two types: commercial-industrial incinerators and domestic incinerators.

Commercial-Industrial Incinerators

Most commercial-industrial waste incineration facilities

use multiple-chamber or controlled air type units. They operate on different principles. Multi-chamber units use high quantities of excess air; controlled air incinerators operate under starved air conditions. The fire hazards associated with each are essentially the same.

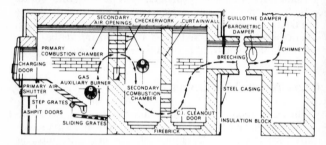

Figure 11-1. A typical commercial-industrial, multiple chamber incinerator.

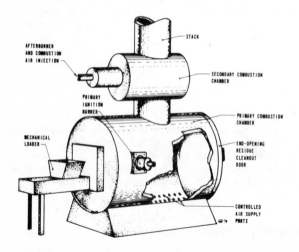

Figure 11-2. Cutaway of a controlled air incinerator.

Incineration Firesafety

The very nature of incineration operations presents potential fire hazards. Incineration firesafety must take into account not only the more obvious measures, such as the presence of sprinklers and room fire rating, but also less apparent, indirect measures relating to facility planning, design operations, and maintenance.

The following items require specific attention for firesafety.

Incinerator and Waste Handling Room: Incinerators and related waste collection and handling equipment must be enclosed within fire-rated rooms or compartments used for no other purpose with self-closing fire doors, suitable for Class B openings.

Equipment Design and Construction: Incinerators, breeching, stacks, and accessories subjected to high temperature combustion reactions and gases should be able to: (1) resist cracking, erosion, warping or distortion; (2) prevent structural failures; and (3) limit outside surface temperatures to not more than 70°F above ambient.

Layout and Arrangement: A good facility provides that: (1) waste and residue containers and the like do not block charging and clean-out operations, or access to work areas; (2) waste material can be charged in a smooth, efficient manner; (3) all parts of the incinerator, including the ash pit, combustion chambers and flues, are easily and safely accessible for cleaning, repair, and servicing; and (3) clearances above the charging door and between the incinerator top and sides to combustible materials meet applicable building codes and the NFPA Incinerator and Waste Handling Standard requirements.

Charging Systems

Charging refers to the feeding or loading of waste materials into the incinerator. An improperly designed or operated charging system could permit flames and combustion products to escape the incinerator and ignite nearby waste materials.

The door opening between the incinerator furnace and outside surrounding areas is the critical element. This opening, normally sealed by a refractory-lined door, must be used frequently for feeding waste materials. Thus, appropriate measures must be taken to assure that heat and combustion products do not escape.

There are basically two types of incineration charging systems — manual and mechanical.

Manual Systems: As the name implies, manpower is used to put waste materials into the incinerator. Manual charging is generally employed on small capacity units. Fire prevention measures are directed primarily towards minimizing the length of time charging doors need to be opened.

Mechanical Charging: Mechanical systems, using mechanical devices to load the incinerator, are generally used on units with capacity in excess of about 500 lb/hr. Mechanical loading devices allow incinerators to be charged with small batches of waste at frequent, regulated time intervals. This protects against over-charging and provides for a continuous and efficient combustion process.

The most common charging system is the "hopper/ram assembly." Waste is loaded into a hopper, a hopper cover closes, a guillotine-type door opens, and a ram then pushes the waste into the incinerator.

Three major fire and smoke problems must be guarded against in mechanical systems. They are:

1. Waste can lodge under the fire door and prevent it from closing tightly. The lodged waste could ignite and spread fire back into the hopper, and possibly beyond. The fire door should be closed by power, not gravity, in order to seal the furnace opening as tightly as possible should any waste become lodged beneath.

2. When the charging ram injects waste into the in-

cinerator, the ram face is directly exposed to furnace heat. Eventually, it will heat up enough to be a potential fire hazard in itself. The charging ram should be cooled either by an internal water circulation system or by a water spray system which quenches the ram face after every charging cycle.

3. When the guillotine door remains open too long, ambient air will surge into the incinerator furnace, upset furnace draft, and result in heavy smoke, and possibly flames, escaping into the charging hopper and operating areas.

The fire door should be interlocked with the hopper cover, or outer door, to prevent opening both at the same time when the incinerator is in operation.

To protect against accidental ignition of waste materials in hoppers, the following should be considered: (1) an ultraviolet flame scanner, (2) an audible alarm signal to indicate hopper fires, (3) a high-temperature automatic sprinkler supplied independent from the room sprinkler system, (4) a manually activated water spray or quench valve, and (5) an emergency switch that would override the normal automatic charging cycle controls and cause immediate injection of hopper contents into the furnace.

Auxiliary Fuel Systems

Burners are provided in almost every incinerator application. They help burn a greater variety of waste as well as providing the important functions of ignition, preheating, and odor and smoke destruction. Burners should be equipped with an electronic flame safeguard system that automatically shuts off burner fuel supply should the burner fail to ignite, have its flame extinguished or encounter insufficient furnace draft. If you find a burner unit without a safeguard system, recommend one.

Domestic Incinerators

Domestic incinerators are classified as gas appliances. The

greatest problem and hazard of domestic incinerators is not
with normal garbage and waste, but with large collections of
highly combustible wastes, such as old floor tiles, scrap rug
remodeling debris, etc. Faced with a big pile of trash to burn,
people sometimes try to stuff too much into the incinerator,
causing exceedingly high heat release and high temperature
in the fire box and chimney connector.

Clearances to combustibles are important. The clearance
above a charging door is 48 inches; clearances from the sides,
rear, and top, 36 inches; and from the front, 48 inches. Side,
rear, top and front clearances may be reduced under special
conditions. Check the clearances carefully and consult the
NFPA Incinerator Standard for permissible variations in
clearances.

WASTE COMPACTORS

Waste compactors use electro-mechanical-hydraulic means
to reduce the volume of waste and to package it in a reduced
condition. Compactors are of two types: commercial-
industrial and domestic.

Commercial-Industrial Compactors

There are four types of compactor systems:

Bulkhead Compactor: Waste is compacted in a chamber
against a bulkhead. When a compacted block is ready for
removal, a bag is installed and filled with the compacted
block.

Extruder: Waste is compacted by being forced through a
cylinder that has a restricted area. Driving forces compact
the material and extrude it into a "slug," which is broken off
and bagged or placed in a container.

Carousel Bag Packer: Waste is compacted into a container
in which a bag has been inserted.

Container Packers: Waste is compacted directly into a bin, cart, or container. When full, the container can be either manually or mechanically removed from the compactor and compaction area.

Chute Termination Bin

Waste chutes to compactors generally do not feed directly into the compactor, but into a small storage chamber, chute connector, or impact area. The potential fire hazard can be minimized by installing sprinklers and providing doors of sufficient size to allow access in the event of fire. If a compactor is charged manually from a large bin with an open top, the bin need not be sprinklered. It is sufficient to rely on the sprinklers protecting the compactor room for protection.

The nature of the bottom closure of a storage bin or area and its ability to be opened under fire conditions deserve attention. If, for instance, there is equipment failure, waste will not only build up into the chute termination chamber, but also into the chute. In the event of a fire and sprinkler discharge, the weight of soggy refuse could become excessive and jam simple slide devices. Sufficient strength should be built into these closure devices to allow opening them without breakage under these conditions.

Compactor Rooms

Compacted material has a wide range of densities and can burn to produce large amounts of smoke; thus, storage of compacted materials in buildings should be minimized. Compactor operations are required to be enclosed with 2-hour fire-rated wall, ceiling, and floor assemblies with fire doors suitable for Class B openings. Automatic sprinklers are also required in compactor rooms.

SHREDDERS

Shredders reduce waste to a uniform size. The possiblity of

explosion is a particular hazard with shredder operations. Explosions can result from ignition of dust-laden air mixtures, which normally surround the shredder during operation. Methods of protection are to provide an explosion suppression system within the shredder room and an explosion vent, preferably above the shredder.

Feed bins to shredders and storage areas for shredded and unshredded waste should be sprinklered. All storage areas should be enclosed within fire-rated rooms with fire doors suitable for Class B openings.

SI UNITS

The following conversion factors are given as a convenience in converting to SI units the English units used in this chapter.

$$1 \text{ in.} = 25.4 \text{ mm}$$
$$1 \text{ lb/hr} = 0.454 \text{ kg/hr}$$
$$\%(°F - 32) = °C$$

BIBILOGRAPHY

McKinnon, G. P. ed., *Fire Protection Handbook*, 15th ed., National Fire Protection Association, Quincy, MA, 1981. Section 7, Chapter 6, Waste Handling Systems and Equipment, discusses the systems and equipment available for handling and disposing of waste so as to minimize its hazard potential.

NFPA Codes, Standards and Recommended Practices. (See the latest *NFPA Codes and Standards Catalog* for availability of current edition of the following document.)

NFPA 82, *Standard for Incinerators, Waste and Linen Handling Systems*. Contains requirements for reducing fire hazards associated with the installation and use of incinerators, waste handling systems, linen handling systems, compactors, and waste storage rooms and containers.

Chapter 12

MATERIALS HANDLING SYSTEMS

Methods of moving materials from one place to another in plants, warehouses, and other occupancies should be of major interest to inspectors. Efficient handling of materials requires highly specialized equipment that can introduce serious fire and explosion hazards unless care is given to their selection, installation, protection, operation, and maintenance. Included are industrial trucks, mechanical stock conveyors, pneumatic conveyors and cranes.

INDUSTRIAL TRUCKS

Industrial trucks are available in many designs and sizes, the most common being the familiar lift trucks of the fork or squeeze-clamp type. Unless these vehicles are properly maintained and used, they can be extremely hazardous.

Industrial trucks may be propelled by electric storage batteries or by gasoline, LP-gas, or diesel engines.

Operation of Trucks

An axiom for the safe use of industrial trucks, particularly in hazardous areas, is using only trucks listed for that type of service. A system of markings has been developed that helps to identify the different types of trucks and the areas where they may be used. The appropriate marking is affixed to both sides of the truck so that its type is easily identified.

Similar to the markings for trucks are markers of corresponding shape, for posting at the entrances to hazardous areas. Thus, when a truck approaches a hazardous area, there's a chance for a quick check, by comparing markings, to ascertain whether the truck should be driven into the area.

The question is: What types of trucks are permitted in

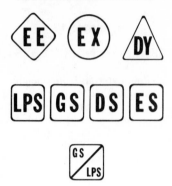

Figure 12-1. Markers for identifying industrial truck types and for posting areas where various types of trucks would be permitted. The signs have black borders and lettering on a yellow background. See the text for definitions of the different types of trucks.

which hazardous location? NFPA 505, Standard for Industrial Trucks, lists thirteen different types of trucks and identifies the hazardous areas where the various types must be used. The alert inspector should be acquainted with the terms of that standard so advice can be given on when the different types of trucks can and cannot be used.

Battery-Powered Trucks

Four types are available for use in locations with hazards ranging from ordinary to severe:

Type E: The minimum necessary safeguards for use in ordinary hazard areas.

Type ES: Additional safeguards to prevent emission of sparks from the electrical system and to limit surface temperatures for areas where easily ignitible fibers are stored or handled (except in process of manufacturing).

Type EE: The electric motor and all other electrical equipment are completely enclosed for use in hazardous locations other than those that require Type EX.

Type EX: Explosion-proof-type (Class 1, Group D) or dust-tight-type (Class II, Group D) construction for areas where there are likely to be explosive mixtures of flammable vapors or combustible dusts during normal operations.

Gasoline-Powered Trucks

Two types are available.

Type G: The minimum necessary safeguards for use in areas of light fire hazard.

Type GS: Additional safeguards in the electrical, fuel, and exhaust systems for occupancies where there are readily ignited combustible materials.

Diesel-Powered Trucks

Three types are available.

Type D: Comparable in hazard to the Type G gasoline-powered truck.

Type DS: Comparable in hazard to the Type GS gasoline-powered truck.

Type DY: Equipped with additional safeguards which make them less hazardous than Type DS gasoline-powered trucks. Surface and exhaust gas temperatures are limited to a maximum of 325 °F, there is no electrical system, and other safeguards are provided in order to minimize the fire hazard normally associated with internal combustion engines.

LP-Gas-Powered Trucks

Two types are available: Type LP and Type LPS. They are considered comparable in fire hazard to Types G and GS gasoline-powered trucks, respectively.

Dual-Fuel Trucks

Two types of dual-fuel trucks are available. They are equipped to be operated on either gasoline or LP-Gas, and they are designated either Type G/LP, units comparable in fire hazard to Types G and LP, or Type GS/LPS, units that require the same safeguards against the hazards of exhaust, fuel, and electrical systems as do Types GS and LPS.

The uniform marking system for identifying the thirteen types and for posting the areas where they can be used is summarized in Figure 12-1. Many fires involving industrial trucks that have spread beyond the truck to involve other property have been the result of using trucks that had no business being there in hazardous locations.

Fire Hazards of Industrial Trucks

The greatest potential fire source for gasoline-, diesel- and LP-Gas- powered trucks are fuel leaks that are ignited by the hot engine, hot muffler, ignition system, other electrical equipment, or other sparks. This danger is somewhat less for diesel trucks because of the higher flash point of diesel fuel; however, it is especially present in LP-Gas trucks as the vapors are difficult to disperse and tend to gravitate toward lower spots or pits. Warn that particular care must be exercised with LP-Gas trucks to avoid the high temperatures near ovens, furnaces, and similar sources of heat.

Battery-powered trucks have experienced comparatively few fires. Nevertheless, electrical short circuits, hot resistors and exploding batteries have been known ignition sources. Frayed wiring and loose terminals are a sign of impending trouble.

Careless and uninformed operation of industrial trucks contributes to property loss. Collision with sprinkler piping, fire doors, and other fire protection equipment; too high tiering or rack storage; and the careless handling of loads, such as containers of flammable liquids, have contributed to many serious fires. If you observe trucks operated in a careless fashion, suggest that perhaps a complete course of instruction for truck operators, or a refresher course at least, would be in order.

Adequate, clear passageways for truck travel and clear warning of overhead and exposed piping will help to reduce accidents. Be particularly sure that sprinkler piping is well marked where there is a danger of elevated loads damaging sprinklers and sprinkler piping.

Maintenance

Good maintenance of trucks is a key to good fire prevention. Observe the trucks carefully for signs of excessive wear and tear and accumulations of grease and dirt. A system of regularly scheduled maintenance based on engine-hour or motor-hour experience can greatly reduce the danger of hazardous malfunctions. Records should show the maintenance that has been performed. Asking to see them is not out of order.

Maintenance and repairs to trucks, particularly to their fuel and ignition systems, should be done only in specially designated areas; they should never be done in hazardous locations. Some points of good maintenance are: filling water mufflers daily or as frequently as is necessary to keep the water supply at 75 percent of the fill capacity; keeping muffler filters clean, and making sure LP-Gas fuel containers are securely mounted to prevent their jarring loose, slipping or rotating. Have removed from service immediately any trucks that are seen giving off sparks or flames from the exhaust system or when the temperature of any part of the truck is found to be in excess of normal operating temperature. Do not permit the trucks to return to service until the faults have been corrected.

Other good practices to observe are keeping the trucks in a clean condition reasonably free of lint, excess oil and grease. Noncombustible agents are preferred for cleaning. Providing each truck with a portable extinguisher suitable for the fuel it uses, e.g., an extinguisher suitable for use on a Class C fire for electrical trucks, is also a good practice.

Refueling and Recharging

Refueling and battery recharging operations should be conducted in well-ventilated areas (outdoors where practicable) away from manufacturing and service areas. Make sure the area is posted for "no smoking," that fuel dispensing pumps are suitable for that use, and that scales for weighing LP-Gas containers are calibrated for accurate filling. Watch the fueling operation. Almost half the fires involving liquid-fueled trucks are the result of spillage during refueling.

Battery recharging operations require special precautions. The corrosive chemical solutions (electrolytes) in the batteries present a chemical hazard. On charge, they give off hydrogen and oxygen which, when combined in certain concentrations, can be explosive. Notice if the recharging area includes means for flushing and neutralizing spilled electrolyte, has a barrier for protecting charging apparatus, and, most importantly, that the ventilation is adequate to carry away fumes from gassing batteries. Make sure that, during the charging operation, the vent caps are kept in place to avoid electrolyte spray. Check to make sure that the vent caps are functioning and that battery covers remain open during charging to dissipate heat.

CONVEYING SYSTEMS

Mechanical conveyors and elevators are among the most commonly used equipment in materials handling. There are many classes and designs for mechanical conveyors including the common belt conveyor, which presents two principal hazards - that of the material being carried and that of the belt itself. And, as is the case with conveyors of any type, there is the danger of communicating fire from one building or area to another.

The materials carried on conveyors range from noncombustible goods (frequently in cardboard containers) to loose fuels and grain.

Fire Causes in Conveyors

Friction between the conveyor belt and a roller or other object is a common cause of fire. Conveyors should be inspected regularly to detect any belt slippage or defective rollers. Moving parts on conveyors should be lubricated on a definite schedule, and inquiries on the inspection frequency and lubricating schedule are not out of order.

Careless cutting or welding operations on the conveyor or its housing is another leading fire cause. Hot metal globules can ignite combustibles on the belt, oil soaked debris around it, or the belt itself. If cutting and welding is observed on conveyors, insist that safe practices be followed in the process (see Chapter 24, Welding and Cutting).

Discharging excessively hot materials from kilns, ovens, or furnaces onto the belt is a fire hazard in some plants. Electrical short circuits, spontaneous ignition, smoking and incendiarism are other causes of conveyor fires.

One of the most frequently occurring factors in fires involving mechanical conveyors is a dusty material, a dusty atmosphere or the dust inevitably created by the materials handling process involving loose combustible materials. Protection for conveying dusty materials centers on adequate choke feeds to conveyors to prevent dust clouds, and, if dust clouds cannot be avoided, enclosed-type conveyors can prevent the dust from escaping.

Static electricity is another hazard of conveyors. Check to make sure all parts of the machines and conveyors are thoroughly bonded and grounded to minimize static discharges. Static electricity can also be controlled by the use of belts made of conductive material, static collectors, or by conductive dressings applied to the belt surface.

Loss experience shows that automatic sprinkler or water spray protection is generally needed for important belt conveyors.

Other Mechanical Conveyors

Chain conveyors equipped with hooks and roller conveyors

are commonly used in assembly lines. Screw conveyors, pan, and bucket conveyors are best for handling hot or molten materials. Belt conveyors, on the other hand, are not generally suitable for handling materials over 150°F.

Bucket elevators, mostly found in bulk processing plants, convey loads vertically, and they are susceptible to the same fire hazards as other mechanical conveyors.

PROTECTION OF CONVEYOR OPENINGS

Protecting openings in walls and floors through which conveyors pass is important to prevent fire spread. Many ingenious methods have been designed for protecting openings for conveyors of different types. They center around water spray protection, fire doors with interlocks or counterweights to stop the flow of material, and even to diverting the conveyor through the roof. It would be well to consult the references at the end of the chapter for information on protection of openings for conveyors.

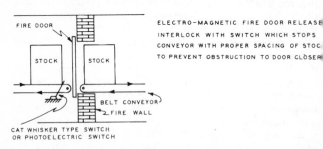

Figure 12-2. Protection of an opening where a belt conveyor can be interrupted at a wall opening.

PNEUMATIC CONVEYORS

Pneumatic systems consist of an enclosed tubing system in which a material is normally transported by a stream of air

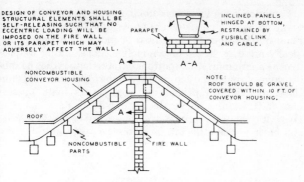

DESIGN OF CONVEYOR AND HOUSING STRUCTURAL ELEMENTS SHALL BE SELF-RELEASING SUCH THAT NO ECCENTRIC LOADING WILL BE IMPOSED ON THE FIRE WALL OR ITS PARAPET WHICH MAY ADVERSELY AFFECT THE WALL.

INCLINED PANELS HINGED AT BOTTOM, RESTRAINED BY FUSIBLE LINK AND CABLE.

PARAPET

A-A

NONCOMBUSTIBLE CONVEYOR HOUSING

NOTE: ROOF SHOULD BE GRAVEL COVERED WITHIN 10 FT. OF CONVEYOR HOUSING.

ROOF

NONCOMBUSTIBLE PARTS

FIRE WALL

Figure 12-3. A conveyor carried over a fire wall.

having a sufficiently high velocity to keep the conveyed material in motion. Sometimes noncombustible gases are used in place of, or mixed with, air, where concentrations of powders and dusts are within the explosive range.

Pneumatic systems are of two principal types: 1) pressure-type which utilizes air at greater than atmospheric pressure and 2) suction-type that transports materials using air at less than atmospheric pressure.

In an air conveying system, any dry collector must be considered as an explosion hazard containing a dust-air mixture in the explosive range. Look at the collector carefully to make sure it is in a safe location with a good barricade or other means of protecting personnel. The collector should be constructed of nonferrous, nonsparking metal or nonmagnetic, nonsparking stainless steel. Where conveying ducts are exposed to weather or moisture, examine them carefully to make sure they are moisture tight. Moisture entering the system can react with dust, generating heat and serving as a potential source of ignition. Likewise, if the conveying gas-air mixture is relatively warm and the dust and collectors are relatively cold, gas temperature may drop below the dew point, causing condensation of moisture. If that be the case, suggest insulating the ducts and collectors or providing a heating system.

Explosion relief for ducts can be provided by antiflashback swing valves or rupture diaphragms extending to the outside. Fans and housings for fans used to move air or inert gases ideally are constructed of conductive, nonsparking stainless steel or aluminum. An important point to note is that no dust is drawn through the fan before it enters the final collector. Note, too, if the fan bearings are equipped with suitable temperature indicating devices wired with an alarm device to give notice of overtemperature. But a word of caution: never approach an operating fan for inspection; wait until it's shut down.

CRANES

Cranes are used principally to move heavy materials about. Cranes that move along rails include overhead traveling, gantry, tower, and bridge cranes. Overhead traveling cranes are used either indoors or outdoors. Gantry, tower, and bridge cranes are mainly for outdoor use. Outdoor cranes are susceptible to damage by high winds. Large cranes that move along rails are generally provided with automatic or manual rail clamps or means of anchorage, such as crane traps, wedges, and cables. Inquiries into the method of anchorage would be in order.

Crane operators' cabs should preferably be of noncombustible construction. Note if they are left free of oily waste, rubbish and other combustibles.

Fire extinguisher protection may be desirable in some operators' cabs. In some large equipment, automatic fixed piping systems may be advisable.

BIBLIOGRAPHY

McKinnon, G. P. ed., *Industrial Fire Hazards Handbook*, 1st ed., National Fire Protection Association, Quincy, MA, 1979. Chapter 44, Materials Handling Systems, discusses in detail various materials handling systems found in industrial plants.

McKinnon, G. P. ed., *Fire Protection Handbook*, 15th ed., National Fire Protection Association, Quincy, MA, 1981. Section 11, Chapter 7, Materials Handling Equipment, provides basic information on industrial trucks, mechanical and pneumatic stock conveying systems and cranes, and discusses the fire and explosion hazards inherent with them as well as the protection of conveyor openings in walls.

Mitchell, D. W., et al, "Fire Hazard of Conveyor Belts," RI 7053, Dec. 1967, USDI Bureau of Mines, Washington, D.C.

NFPA Codes, Standards and Recommended Practices. (See the latest *NFPA Codes and Standards Catalog* for availability of current editions of the following documents.)

NFPA 66, *Pneumatic Conveying Systems for Handling Flour, Grain and Other Agricultural Dusts*. A standard on pressure-type and suction-type systems. It contains guidance on design safeguards of component parts of pneumatic conveying equipment.

NFPA 80, *Standard for Fire Doors and Windows*. A standard on the use, installation, and maintenance of fire doors, windows, glass blocks and shutters.

NFPA 505, *Firesafety Standard for Powered Industrial Trucks Including Type Designations, Areas of Use, Maintenance and Operation*. A standard containing definitions of all types of industrial trucks and detailed descriptions of the different hazardous areas where each type can be used.

Chapter 13

HOUSEKEEPING PRACTICES

Good housekeeping is plain common sense. You do not need intensive training to recognize, almost intuitively at first glance, whether or not the housekeeping on the premises being inspected is satisfactory. Cleanliness and orderliness are basic to good firesafety, and if early on the inspector feels a bit uneasy about the quality of the housekeeping, or the general care and management of the property, then eyes had better be opened a little wider in looking over the points of hazard management.

Good housekeeping practices — both indoors and outdoors — are a good method of controlling the presence of unwanted fuels, obstructions, and sources of ignition. Certain aspects of housekeeping are a common denominator to most properties whatever their use; others are peculiar to a particular occupancy. It's neither practical nor possible to describe every feature of housekeeping for all occupancies; the alert inspector will visualize hazardous housekeeping situations peculiar to the occupancy being inspected and be prepared to offer recommendations to eliminate them.

PRINCIPLES OF GOOD HOUSEKEEPING

The basic requirements of good housekeeping fall into three categories:
1. Proper layout and equipment.
2. Correct materials handling and storage.
3. Cleanliness and orderliness.

When proper attention is given to establishing the routines for these three factors, good housekeeping is almost a certainty.

Layout and Equipment

A close look at working areas, storage areas and the equipment used to move materials about should give you a pretty good idea if housekeeping is a problem. Aisles clogged with materials waiting to be processed, for example, can discourage effective overall cleanliness. Maybe some simple rearrangements can improve the housekeeping considerably. At the least, suggest that management perhaps might like to take a look at its workflow procedures if haphazard arrangement adds to the cluttered appearance.

Materials Handling and Storage

Lack of facilities to store materials neatly and to move them about easily compounds the housekeeping problem. Exitways can become blocked; access to fire extinguishers, control valves for extinguishing systems and small hose stations can be blocked and other fire protection equipment, such as fire doors, made inoperative. Disordered storage attracts debris and trash in forgotten corners.

Cleanliness and Orderliness:

The level of firesafety is immeasurably improved where attention is paid to keeping all areas as clean and neat as possible. The principal defense against unsightly and dangerous accumulations of unwanted materials and trash is an efficient and timely waste removal program backed up by each individual's personal sense of responsibility and desire to keep one's surroundings neat and clean. Observe the adequacy of waste removal devices. Are enough noncombustible wastebaskets, bins, cans, and other proper containers provided so that occupants will find tidiness convenient?

CONTROL OF SMOKING

This is frequently a tough problem to solve. The urge to smoke often conflicts with the effect smoking will have on the

Figure 13-1. Waste containers designed to snuff out accidental fires in contents and to limit external container surface temperatures to no more than 175°F above room temperature.

level of firesafety. In some instances complete prohibition is unrealistic; often careful regulation of smoking can achieve the same results.

A 10-point guideline that may prove useful is:

1. Provide safe, supervised and convenient smoking areas, preferably with sprinkler protection;

2. Equip smoking areas with ample smoking receptacles and fire extinguishers; keep the area clean and free of combustibles;

3. Maintain a continuous campaign against reckless smoking and careless use of matches;

4. Designate areas where smoking is prohibited;

5. Allow only safety matches or electrical or mechanical lighters;

6. Crush all butts — killing the glow, and discard them in a safe place;

7. Enforce the "No Smoking" rules in restricted areas;

8. Encourage safe personal smoking habits among employees;

9. See that visitors are informed of the "No Smoking" rule; and

10. Make sure all outside contractors observe these regulations.

Areas where smoking is permitted as well as those where it is limited or prohibited must be clearly marked. Even though the premises are clearly marked as to permitted and prohibited smoking areas, don't be lulled into thinking the problem is entirely solved. Sneaking smokes in remote and congested areas, such as warehouses, is not unknown. Keep an eye out for tell-tale evidence of crushed smoking materials underfoot. Tell management of your findings, and remind them of what could happen. Every little bit helps.

Figure 13-2. Examples of signs permitting or forbidding smoking in designated areas. (Factory Mutual System)

PROCESS HOUSEKEEPING PROBLEMS

Different occupancies have special housekeeping problems depending on the nature of the processes involved. Some of the more common of these special problems are:

Flammable Liquid Spills

This can happen whenever flammable liquids are handled or used. Inquire if there is a supply of suitable absorptive material and tools for spreading it to contain spills. You could review with occupants steps that they could take to cut off sources of ignition, ventilate the area and safely dissipate any flammable vapors from spills.

Flammable Liquid Waste Disposal

This often is a troublesome problem. Waste liquids are never drained into sewers. Good practice is to place them in metal drums until safely disposed of. There are companies that make a specialty of collecting waste petroleum products. Review with management the procedures followed to dispose of flammable liquids.

Coatings and Lubricants

Paints, grease, and similar combustibles are sources of combustible residues. Cleaning spray booths, exhaust fan blades, and ducts must be done frequently to avoid dangerous accumulations. Be particularly alert to the protection given to sprinklers in spray booths. A thin coating of grease placed on sprinklers and cleaned frequently is one method. Another is to enclose each sprinkler in a light plastic or paper bag which is changed daily.

Clean Waste and Rags

Clean cotton waste or wiping rags are considered to be mildly hazardous, chiefly because there is always the likelihood that dirty waste may become mixed with them. Dirty waste or small amounts of certain oils may lead to spontaneous heating. Clean waste is best stored in metal or metal-lined bins with covers, while local supplies of clean waste should be left in small waste cans.

Oily Waste

Oily wiping rags, sawdust, lint, clothing, etc., particularly if containing oils subject to spontaneous heating, are dangerous. Recommend that small amounts be stored in standard waste cans; large amounts in heavy barrels with covers. Suggest that cans containing oily waste be emptied daily.

Figure 13-3. A portable metal waste can equipped with a self-closing cover for storage of oily waste. (The Protectoseal Co.)

Packing Materials

Most packing materials are combustible and extremely hazardous. Styrofoam pellets and shapes, excelsior, straw, and the like should be treated as clean waste except that, where large quantities are used, special vaults or storerooms are frequently needed.

Notice if there are provisions for disposing of used or waste packing materials and crating from shipping and receiving rooms. Large amounts of accumulated waste strewn over the floor around unpacking operations is an invitation to disaster. You can't be too strong in recommendations to clean up the premises.

Lockers and Cupboards

Personal lockers, particularly in industrial plants, can be fire hazards. Lack of cleanliness and the general use of lockers for storing waste materials present dangers. Pipes and cigars that are not extinguished and find their way into lockers are dangerous. Obviously, you might not be able to open personal lockers looking for hazards, but general observations of the locker area will give you a good idea of the standards of cleanliness. If things look bad (doors open with oily clothes thrown haphazardly into the lockers, waste rags on the floor, etc.) you should tell management they have a problem.

BUILDING CARE AND MAINTENANCE

Certain maintenance procedures common to most occupancies are worthy of mention.

Cleaning and Treatment of Floors

Treatment, cleaning, and refinishing of floors can be hazardous if flammable solvents or finishes are used or if combustible residues from refinishing are produced in quantity. In general, cleaning or finishing compounds containing solvents having flash points below room temperature are too dangerous to use except in small quantities. Inquire about the solvents and compounds used in caring for floors. The labels on the containers probably will give information on their contents' flammability.

Floor Oils

Oils and low flash point compounds are a hazard, particularly when they are first applied. Some floor dressings may heat spontaneously. If you find that oiling floors is an activity on the premises, check to make sure that oily mops and rags used in the activity are safely stored in metal or other noncombustible containers.

Flammable Cleaning Solvents

Availability of nonhazardous cleaning solvents which have high flash points, stability, and low toxicity make use of hazardous flammable solvents the rarity rather than the practice. However, flammable solvents may be present and inquiries may reveal that indeed they are being used. In that case make sure that safety cans with tight fitting caps for dispensing small quantities are used. Open pails, buckets and dip tanks without self-closing lids are not permissible.

Kitchen Exhaust Ducts and Equipment

Grease condensating in exhaust ducts from hoods over restaurant ranges is a serious problem. Grease can be ignited by sparks from the range or, more often, by a small fire in cooking oil or fat in pans or deep fryers.

The exhaust system should be inspected daily or weekly depending on how much it is used. Hoods, grease removal devices, fans, ducts and associated equipment all need cleaning at frequent intervals. Make sure flammable solvents or cleaning aids aren't used in the cleaning process.

Be particularly concerned with the inside of the ducts. Cleaning of ducts is often neglected because it's a nasty job. There are commercial firms that do that type of work, and the yellow pages may list some. In any case, never suggest burning the grease out even though ducts installed according to NFPA standards are designed to withstand burnout.

Cleaning

If cleaning is done without commercial assistance, suggest a powder compound consisting of one part calcium hydroxide and two parts calcium carbonate. This compound saponifies the grease or oily sludge, thus making it easier to remove and clean. Another method is to loosen the grease and scrape the residue out of the duct.

Spraying duct interiors with hydrated lime after cleaning is a fire prevention measure. It tends to saponify the grease and make subsequent cleaning easier.

Dust and Lint Removal

Combustible dust and lint accumulation on walls, ceilings, and exposed structural members is a problem in some occupancies. Removing it can be an explosion hazard if it's not done correctly. Using vacuum cleaning equipment having dust-ignition-proof motors is a safe method. In any event, care should be taken not to dislodge into the atmosphere quantities of combustible dust or lint that could form an explosive mixture with air, ignite, and explode.

Caution against blowing down dust with compressed air. It can create dangerous dust clouds and is a method used only as a last resort, and then only after all possible sources of ignition have been eliminated.

OUTDOOR HOUSEKEEPING

Pay as much attention to the grounds outside the building as you do the interior. Poor housekeeping on the outside can threaten the fire security of exposed structures, goods stored in the yard, and the building itself. Accumulations of rubbish and waste and tall grass and weeds growing close to buildings and storage piles are probably the most common hazards. Before entering the premises look around outside for a "feel" for the quality of grounds and yard maintenance. You may want to spend considerable time outside later if your first observations make you suspect problems.

Weeds and Grass

Dry weeds and grass can be controlled by herbicides. Among chemicals used are chlorate compounds, particularly sodium chlorate. A caution: chlorate compounds are oxidizing agents and can contribute to fire conditions, particularly during long hot periods in the summer when the weeds and grasses they have killed dry out.

Calcium chloride and agricultural borax, applied dry or in solutions, are effective and nonhazardous weed killers. Other herbicides are ammonium sulfamate and other commercial chemical weed killers that have little or no fire hazard, and sodium arsenite and other arsenic compounds that are effective herbicides but toxic.

Burning dry grass and weeds to remove them can lead to disaster unless it's controlled burning at the proper time of the year and supervised by the fire department.

Refuse and Rubbish Disposal

Goods stored outdoors require passageways between storage piles that are kept clear of combustibles. Make sure that the aisles are left clear and that separations are maintained between piles of combustibles in storage, and between the piles and nearby combustible buildings. Don't let discarded crates, utility shacks, or other combustible materials clutter the clear spaces. Observe if large receptacles are available for disposal of smoking materials before entering designated "no smoking" areas.

Combustible waste materials from industrial operations often are stored in the yard before being hauled away. Make sure the accumulations of waste are no closer than 20 feet from buildings (50 feet is better). Fences should surround the waste in storage. Ignition sources, such as incinerators, should be kept a safe distance away.

BIBLIOGRAPHY

McKinnon, G. P. ed., *Industrial Fire Hazards Handbook*, 1st ed., National Fire Protection Association, Quincy, MA, 1979. Chapter 42, Industrial Housekeeping Practices, pp. 821-834. Contains information on housekeeping practices in industrial plants.

McKinnon, G. P. ed., *Fire Protection Handbook*, 15th ed., National Fire Protection Association, Quincy, MA, 1981. Section 11, Chapter 8, Housekeeping Practices, provides guidance on general housekeeping practices and control of housekeeping hazards.

"How to Apply Good Housekeeping," *The Handbook of Property Conservation*, Factory Mutual System, Norwood, MA, 1973, pp. 189-192. Principles of good housekeeping are explained.

NFPA Codes, Standards, and Recommended Practices. (See the latest *NFPA Codes and Standards Catalog* for availability of the current edition of the following document.)

NFPA 96, *Standard for the Installation of Equipment for the Removal of Smoke and Grease-Laden Vapors from Commercial Cooking Equipment*. Contains guidance on the installation and maintenance of all the components of exhaust systems for commercial cooking equipment including hoods, grease removal devices, exhaust ducts, dampers, air moving devices, auxiliary equipment and fire extinguishing equipment.

Chapter 14

GENERAL STORAGE

Good storage practices are based on recognizing two important considerations: a) the fire behavior of the commodities in storage, and b) their storage arrangement. Together with the construction of the building, they determine the degree of fire protection required (generally automatic sprinklers backed up by hose streams).

Fire behavior depends on the ease of ignition, rate of fire spread, and rate of heat release of the commodities. Storage arrangements include height of storage, aisle widths, and whether storage is in bulk, solid piles, or palletized piles. Fire development depends mostly on the combustible surfaces of the stored goods and on the horizontal and vertical flue spaces between the surfaces of the goods. Fire in stored goods usually spreads in a fan-shaped pattern; fire at the base preheats the material above, which ignites and burns, increasing in size and intensity as it moves upward rapidly.

Smoking, industrial trucks, sparks from welding and cutting, and arson are the more common causes of fire in storage occupancies. Regardless of the actual ignition source, bad housekeeping is an almost essential partner in the ignition process. Packing and unpacking of goods require quantities of loose combustible substances, such as polystyrene beads, cocoons of foamed plastic, shredded paper, etc. Evidence of sloppiness in handling packing materials should be a clue that housekeeping may be a major problem. Scrap materials allowed to accumulate can easily be ignited and act as kindling for stored items. Good housekeeping cannot be overemphasized.

STORAGE ARRANGEMENTS

For the purpose of fire protection, consider storage as divided into four main categories: bulk storage, solid piling, palletized storage and rack storage. (Bins and narrow shelves are also important, but they are usually found in stockrooms and mostly hold only small amounts of materials for immediate use). The chief difference between the four major categories having an effect on fire behavior and fire control is the nature of horizontal and vertical air spaces or "flues" that the storage configurations create.

Bulk Storage

Powders, granules, pellets, and flakes, all in free flowing condition, are the principal forms of materials for bulk storage. Silos, bins, tanks, and large piles on the floor are the storage methods. Fire in large piles tends to burrow into the piles. Spontaneous ignition fires can start in the interior and may be difficult to locate and extinguish.

Pay close attention to bulk storage. Learn about the fire characteristics of the materials you find in bulk storage. Automatic sprinkler protection often is not effective on the burrowing fires that may occur. Sometimes the fire can be put out only by removing the burning material from the storage facility. Are there provisions for quick removal of bulk commodities from the building? Find out what management's game plan is in an emergency in bulk storage.

Conveyor equipment, such as belt conveyors, air fluidizing through ducts, and bucket conveyors agitate materials as they move to and from bulk storage. If combustible dust clouds are generated, notably in grain storage facilities, there can be an explosion hazard. Automatic sprinklers are frequently needed in the housings around conveyor equipment.

Solid Piling

Cartons, boxes, bales, bags, etc., in direct contact with each other to the full dimension of each pile make solid piles. Air spaces, or flues, exist only where there is imperfect contact or where a pile is close to, but not touching, another pile.

Compared with palletized and rack storage, solid piling gives fire the least opportunity for development; however, if outer surfaces possess rapid flame spread properties, high solid piles can be a severe hazard.

Palletized Storage

Palletized storage consists of unit loads mounted on their pallets that can be stacked one on the other. Each pallet is about 4 inches high, usually of wood, but some are made of metal, plastic, expanded plastic, or cardboard. The height of palletized storage is limited by the "stackability" of a commodity (its resistance to crushing at the lowest part of the pile), which is usually 30 feet at the most.

The severe hazard of palletized storage is the horizontal air spaces formed by the pallets themselves within each layer of pallets. These spaces, more often than not, are continuous in one direction for the entire width of the pile, creating a long horizontal flue mostly out of reach of water from sprinklers.

Figure 14-1. A conventional wood pallet.

Idle Pallets

Piles of wood or plastic pallets are a severe fire hazard. After a short time of use, pallets can dry out and their edges become frayed and splintered. They can ignite easily, and, even with sprinklers operating, the underside of the pallets provide a dry area where fire can grow and expand. Normally 6 feet is the maximum height for idle pallets in storage to keep fire within limits manageable by automatic sprinklers discharging at ordinary densities (0.20 gpm/sq ft); higher piles require higher densities. If there is no sprinkler protection, suggest strongly that unused pallets should be stored outdoors.

Rack Storage

Rack storage consists of a structural framework that can hold unit loads, generally on pallets. The height of storage racks is limited, potentially, only by the vertical reach of the materials handling machinery which, like the racks themselves, can be designed for great heights. Generally, racks are about 25 feet high. In fully automatic warehouses, racks are sometimes up to 100 feet and have even exceeded that height. In some installations the steel storage racks are adapted as part of the building framing.

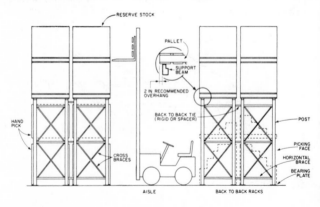

Figure 14-2. A common arrangement of double row racks with palletized storage atop. (Unarco Storage)

As in low palletized storage, longitudinal and transverse flue spaces in high rack storage are significant in fire spread and sprinkler fire control. Transverse flue spaces occur at the uprights and between positions of pallet loads.

Figure 14-3. High steel storage racks as structural support for roofs and walls of a storage facility. (Clark Handling Systems Div.)

FIRE PROTECTION FOR GOODS IN STORAGE

Automatic Sprinklers

Good automatic sprinkler protection is the backbone for protecting goods in storage, whether it be low solid piles or super-high rack storage with in-rack sprinklers. But providing good sprinkler protection is a bit more complicated than applying a single set of installation rules to all sprinkler installations. Sprinkler system and waterflow requirements for goods in storage generally are calculated on the basis of a general hazard classification system for commodities and sprinkler system design curves for storage of the different classes of commodities.

The commodity classification system is found in NFPA 231, *Standard for Indoor General Storage*, as well as in NFPA 231C, *Rack Storage of Materials*. The system establishes four categories (Class I, II, III, and IV) with slight variations between the requirements for rack and general storage (the latter recognizes plastics as a special problem). These four classifications were established to classify commodities according to the difficulty of fire control by automatic sprinklers; Class I is the easiest to control. Different sprinkler design curves apply to the different classes of commodities. These design curves will be found in NFPA 231 and NFPA 231C as well as in NFPA 231D, *Standard for Storage of Rubber Tires*.

It is beyond the scope of this manual to cover the use of the classification system and the design curves. Consult the three storage standards previously mentioned and NFPA 13, *Installation of Sprinkler Systems*, for the detailed information you will need to judge the adequacy of a sprinkler installation in a storage occupancy.

One important point to check, though, is clear space between sprinklers and the tops of storage piles. A minimum of 18 inches is required so that water distribution is not hindered (modifications of clearance requirements for rack storage are given in the NFPA Rack Storage Standard). Obstructed sprinklers are almost worthless.

Hydrants, Hose Streams, and Extinguishers

Hydrants: Small storage facilities close to public hydrants present no special problem, but, in larger buildings where the small dimension exceeds about 200 feet, hose lays from public hydrants to the far side of the building can be a serious problem. Private hydrants are needed where public hydrants do not give adequate coverage. Locating the hydrants near points of entrance to the building cuts down on the amount of hose needed.

Small Hose: Stations for 1½-inch hose should be located throughout the premises so that all areas can be reached. The hose can be supplied from a wet-pipe sprinkler system or a separate piping system.

Portable Fire Extinguishers: Extinguishers are needed for Class B (flammable liquid) and Class C (electrical) fires and may even have use on Class A (ordinary combustible) fires even though small hoses are available. Class B extinguishers often are mounted on in-plant vehicles. Even though portable extinguishers are available, they may have only limited effectiveness on piles; often the piles have to be pulled apart manually to extinguish fires and hand lines used for control during mop-up operations.

STORAGE OF SPECIFIC MATERIALS

Some commodities deserve special mention because of the nature of the fire hazards involved.

Rubber Tires

Tires in storage are rated as a very high fire hazard. They burn readily, releasing much heat and smoke.

Tires are stored directly on the floor, either on their sides or in pyramided piles, or in fixed or portable racks. They may also be stored where a number of tires are strapped together in palletized portable racks.

Automatic sprinklers with adequate discharge densities can control a fire, but cannot be expected to put it out. Thus, it is essential that means for venting the building to remove smoke are provided to permit prompt entry for manual fire fighting and overhaul after the fire has been brought under control by sprinklers.

Roll Paper

Roll paper is stored on its side, nested between rolls of a lower tier or resting on dunnage placed between tiers, or is stacked on end. Rolls may also be stored in racks. The most hazardous configuration and the greatest challenge to sprinkler protection is presented by rolls stored on end as separate columns.

Fire fighting is best handled by automatic wet-pipe sprinkler systems. Hose streams must be carefully managed so that they do not rob the sprinklers of water, which might cause the loss of fire control. High rates of water discharge for long durations are necessary to supply sprinklers, particularly for the protection of large quantities of storage or high-piled storage.

Storage of Plastics

The fire hazard of plastic materials in storage is determined by their chemcial composition, their physical form, and the storage arrangement. The physical form may be foam, solid, sheet, pellets, flakes, or random small manufactured objects in bags or cartons.

Plastics, such as fluorocarbons, unplasticized polyvinyl chloride, and phenolics, present about the same fire hazard as general storage commodities. Pellets and small thermoplastics, such as polyurethane, polyethylene, and plasticized polyvinyl chloride, and thermosets, such as polyester, present a severe fire hazard exceeded only by thermoplastics, such as polystyrene and acrylonitrile-butadienestyrene (ABS). These plastic materials will melt and break down into their monomers and burn like a flammable liquid. High sprinkler discharge densities over large areas are necessary to protect them. When in the form of a foamed material, these plastics present the most severe fire hazard, requiring the greatest sprinkler discharge density and area of operation.

Refrigerated Storage

Temperatures in cold storage warehouses range from 32°F to 65°F for products such as fruits, eggs, or nuts that would be damaged by freezing down to 0°F to −35°F for initial freezing. Whether the building construction is combustible or not, insulating materials generally are combustible—for example, corkboard or expanded (foamed) plastics. Widely used insulating materials are foamed polystyrene and polyurethane, which melt at a low temperature, burn rapidly, and release large quantities of smoke. A coating of portland cement plaster over the explosed plastic surfaces will prevent melting and ignition of the plastic.

Combustible materials found in cold storage warehouses include wood dunnage, wood pallets, wood boxes, fiberboard containers, wood baskets, waxed paper, heavy paper wrappings, and cloth wrappings. There generally are sufficient quantities of these combustibles present to have a fire severe enough to require sprinkler protection. Sprinkler systems may be either preaction or dry-pipe, with the former preferred.

Hanging Garments

Garments stored on pipe racks are vulnerable to smoke and water damage. They may extend to a height of two or three tiers. They may be of cotton, wool, synthetic fabric, or blends. Different types of fabrics burn at different rates, but the prevalence of blends tends to equalize the nature of the fire problem. Garments may be stored with or without plastic dust covers (which have no appreciable effect on their burning).

Carpets

Rolls of carpeting, commonly 12 to 15 feet long, are stored in deep shelving called racks that are sometimes back-to-back, so that distances between aisles may total 30 feet. The rolls may be individually stored in strong cardboard tubes, on solid or slatted shelves, or in racks arranged in cubicles. Each tier (shelf to shelf) is only about 2 to 3 feet high, with as many as ten tiers in a rack. Racks frequently run 100 feet or more in length.

The type of storage described does not permit much access of water from ceiling sprinklers, so that in-rack sprinklers are frequently necessary. A method of slowing the longitudinal spread of fire is to provide a few inches of open space at vertical supports; another is to provide vertical transverse barriers at every third set of vertical supports.

OUTDOOR STORAGE

Outdoor storage requires special protection. There are so many different things to consider that no set rules apply to say what exactly constitutes that protection. The best that can be done is to outline general principles and rely on the experience and judgment of persons who will apply them. NFPA 231A, *Recommended Safe Practice for Outdoor General Storage*, can provide you with that guidance.

In general, outdoor storage sites should be level and firm underfoot with adequate clearances so that fire can't spread to the site from other sources. Areas where flooding and windstorms are problems should be avoided.

Some general principles also apply to site layout. The density of packed materials makes a great difference in the way they burn. Lumber stacked solid does not burn as rapidly as lumber in "sticked" piles with air spaces between boards. Some materials burn quickly and produce a great deal of heat. Baled hay and cotton, lumber, packing materials, pallets, pulpwood and rubber are examples.

Access to the yard and the piles in it must be made easy.

Driveways at least 15 feet wide permit fire apparatus to reach all portions of the yard. Aisles should be at least 10 feet wide and, for unusually large storage areas or moderate-sized yards with valuable commodities, main aisles or firebreaks can be used to subdivide the storage. The actual width of aisles can be a matter of judgment depending on the combustibility of the commodity, how it is stored, height of piles, nearness to buildings, wind conditions, fire fighting forces available, etc.

Piles of materials stable under normal conditions may collapse in fire and cause severe fire spread, particularly from flying brands.

Adequate public fire and police protection or equivalent private fire brigade protection is a prime requirement for outside storage facilities. Make sure there are an adequate number of gates in the fence surrounding the storage site. If investigation of the flow from public hydrants shows that water flows are inadequate, it may be necessary to suggest private water storage facilities or pumps, or both.

For some storage facilities, such as for lumber or wood chips, and where strong water supplies are available, it may be practical to use monitor nozzles mounted on towers. Appropriate portable fire extinguishers placed at well-marked strategic points throughout the storage area are practical. If people are usually present, fully equipped hose houses would be welcome additions providing personnel are trained in using hose lines.

Another important feature to check on is the availability in the area of a means to notify the fire department. At the least, make sure there's a telephone handy.

IDENTIFICATION OF MATERIALS

It's fairly easy to identify fire hazards of commonly found materials in storage, such as wood, paper, fabrics, and LP-Gas cylinders. You know pretty well what to expect when they burn. But there are literally thousands of combustible solids, flammable and combustible liquids, and liquefied and compressed gases for which the hazards are not so readily apparent. They need a system for identifying the hazards so that the right response can be taken in emergencies involving them. Chief among the hazard identification systems is the NFPA 704 System of Hazard Identification.

The NFPA 704 System

The NFPA 704 information system is based on the "704 diamond," which is a design for visually presenting information on health, flammability, and self-reactivity hazards plus special information associated with the hazards at fixed installations such as storgae rooms and warehouses. The system was not designed to be used for materials in transportation.

Figure 14-4. The NFPA 704 Identification System Diamond.

Numbers from 0 to 4 are placed in the three upper squares of the diamond for each of the three hazards. The "0" indicates the lowest degree of hazard; the "4" the highest. The fourth square at the bottom is used for special information. Two symbols for this latter use are: a "W" with a bar through it which warns *not* to use water indiscriminantly and the "radioactive pinwheel," the familiar symbol for radioactive hazards.

The five degrees of hazard, in the order of their descendency, are:

4—Too dangerous to approach with standard fire fighting equipment and procedures. Withdraw and obtain expert advice on how to handle.

3—Fire can be fought using methods intended for extremely hazardous situations, such as unmanned monitors or personal protective equipment which prevents all bodily contact.

2—Fire can be fought with standard procedures, but hazards are present which require certain equipment or procedures to handle safely.

1—Nuisance hazards present which require some care, but standard fire fighting procedures can be used.

0—No special hazards, therefore no special measures.

The preceding gives the general meanings of the degree of hazard each number represents. More details on the three categories of hazards and hazard levels which the various numbers indicate are given in the text and appendices of NFPA 704, *System for the Identification of the Fire Hazards of Materials*, referenced at the end of this chapter.

Using the 704 Diamond

The "diamond system" can be used in many ways. Considerable leeway is permitted in the presentation of the numbers. The only basic requirement is that numbers be spaced as though they were in the diamond outline. Sometimes color (blue for health hazard, red for flammability, and yellow for self-reactivity) is used for solid backgrounds in the diamond or to color the appropriate numbers themselves. Stencils, labels, and placards are some of the forms the system can take. They will be found on the sides of tanks, drums, barrels, boxes, on chemical processed equipment, entrances to laboratories, or wherever a warning is needed of the presence of hazardous materials. Look for the symbols; perhaps further uses of the system can be suggested.

Guidance for assigning hazard identification numbers to materials can be found in NFPA 325M, *Fire Hazard Properties of Flammable Liquids, Gases and Volatile Solids*, and NFPA 49, *Hazardous Chemicals Data*. The effects of local conditions must be considered. For instance, a drum of carbon tetrachloride sitting in a well-ventilated storage shed presents a different hazard from a drum sitting in an unventilated basement.

STORAGE OF RECORDS

Good records are the life blood of a successful organization, and the measures used to protect valuable records from flame, heat, smoke, and exposure to water are important to an organization's survival. But the exact measures used for protection of records should be based on a survey to ensure that the proper records are selected for special protection. Not all records deserve the same degree of protection.

Records can generally be classified in one of two ways for purposes of assessing their value: 1) vital records that are irreplaceable, e.g., records that give direct evidence of legal status and ownership, and which are needed to sustain a business; and 2) important records that can be reproduced from original sources only at considerable expense.

Other records may be useful, but should be kept well separated from vital and important records, because they may constitute a fire exposure themselves. Bulk storage of less important paper records in file cabinets, various arrangements of shelving, palletized cardboard boxes, etc., present enough combustible material to cause burnout and severely damage the building in which it is located. Photographic films and magnetic tapes having an acetate or polyester base have the same order of combustibility as paper, but tapes on reels and in plastic cases produce abnormal amounts of heat, smoke, and toxic gases.

Automatic sprinklers are the preferred form of fire protection for limiting the loss in bulk quantities of records. Special extinguishing systems (foam, carbon dioxide and halon agents) are useful in special situations. If you find bulk storage of records, make sure that the protection provided is appropriate and in good working order, and the obvious hazards and exposures have not been overlooked.

Vaults

The term "vault" refers to a completely fire resistive enclosure up to 5,000 cubic feet in volume used exclusively for records storage. Vaults usually contain a substantial fuel load and by their nature contain only vital and important records. In many instances, the contents of some vaults are more of a hazard than any external fire exposure. A fire within an un-protected vault can be disastrous unless it can be discovered immediately and extinguished.

Formerly, no wall penetrations in vaults were permitted except for the door opening. This prohibited sprinklers, fire and smoke detection units wired to a master panel, and even fixed lighting systems. Now vault standards permit wall penetrations for piping and conduits so vault contents can be given better protection. If you find older vaults without inter-nal detection and protection devices, you should suggest their installation. In any event, water-type fire extinguishers or small hose, or both, are desirable in an accessible location near the door of the vault.

File Rooms

File rooms are built as nearly like vaults as possible, but they are for situations in which people work regularly with the records in the room. They may have electric lights and steam or hot water heat. Wall openings, if needed for air con-ditioning or ventilation, must be equipped with fire dampers. Standard file rooms have a maximum ceiling height of 12 feet and a maximum volume of 50,000 cubic feet. File room doors may have vault door ratings or lesser ratings of ½ or 1 hour. Automatic sprinklers are desirable.

Archives and Record Centers

Bulk storage of paper records in separate buildings, in a major portion of a building, or in a room exceeding 50,000 cubic feet in volume requires special attention because of the large fire load it represents. The four basic factors that must be considered are (1) exposure from nearby operations or buildings, (2) the potential for ignition, (3) the potential for fire spread, and (4) the capability of the available fire control system to extinguish or control the fire with minimum damage to records.

Fire resistive construction is essential to protection against exposure fires. Cleanliness, orderliness, and an absolute ban on smoking are the fundamentals of controlling the chances of ignition. The type of storage (steel cabinets or open shelf system) governs the potential for fire spread. Open shelves present a wall of paper at the face of the shelves, up which fire can spread rapidly to involve other rows of shelves.

Automatic sprinklers, backed up by an early warning detection system, provide good protection in view of the rapidity with which fire can spread in the large open areas customarily found for archival storage.

Safes and Record Containers

Safes and containers are available with typical ratings of 4, 2, 1, and ½ hours. There are two types of containers and safes — Class 150 principally for the storage of magnetic tapes, and Class 350 for the storage of paper records. Insulated filing devices are available with a Class 350, ½-hour rating.

SI UNITS

The following conversion factors are given as a convenience in converting to SI units the English units used in this chapter.

$$1 \text{ in.} = 25.4 \text{ mm}$$
$$1 \text{ ft} = 0.305 \text{ m}$$
$$\tfrac{5}{9}(°F - 32) = °C$$
$$1 \text{ cu ft} = 0.0283 \text{ m}^3$$
$$1 \text{ gpm/sq ft} = 40.746 \text{ litres/min/m}^2$$

BIBLIOGRAPHY

McKinnon, G.P. ed., *Fire Protection Handbook*, 15th ed., National Fire Protection Association, Quincy, MA, 1981. Section 4, Chapter 11, Identification of the Hazards of Materials, discusses in detail the various systems that are available to identify the hazards of materials to minimize danger to emergency personnel; and Section 10, Chapter 1, General Indoor Storage (Without Racks); Chapter 2, Indoor Rack Storage; Chapter 4, Outdoor Storage Practices; and Chapter 10, Protection of Records, all give guidance on storage practices, facilities, and fire protection for goods and records in storage.

McKinnon, G.P. ed., *Industrial Fire Hazards Handbook*, 1st. ed., National Fire Protection Association, Quincy, MA, 1979. Chapter 41, Industrial Storage Practices, discusses storage practices observed in industrial locations including special storage facilities (piers and wharves, underground storage, air-supported structures, etc.)

NFPA Codes, Standards, and Recommended Practices. (See the latest *NFPA Codes and Standards Catalog* for availability of current editions of the following documents.)

NFPA 231, *Standard for Indoor General Storage*. Contains guidance for the storage of combustible commodities in buildings with automatic sprinkler systems. Emphasis is on providing adequate sprinkler discharge for goods of different degrees of combustibility.

NFPA 231A, *Outdoor General Storage*. Contains recommendations for storage, handling, and safeguarding of commodities stored outdoors.

NFPA 231C, *Standard for Rack Storage of Materials*. A standard applying to the storage of a broad range of combustible commodities in racks over 12 feet high.

NFPA 231D, *Standard for Storage of Rubber Tires*. Covers storage arrangements and fire protection requirements for large quantities of rubber tires.

NFPA 232, *Standard for the Protection of Records*. Contains requirements for records protection equipment, facilities and records handling techniques.

NFPA 232AM, *Manual for Fire Protection for Archives and Record Centers*. Gives guidance on fire protection for file rooms exceeding 50,000 cubic feet and for all archives and record centers.

NFPA 325M, *Fire Hazard Properties of Flammable Liquids, Gases, and Volatile Solids*. A tabulation of available data on the fire hazards of more than 1,500 substances including suggested hazard identification numbers for use with the NFPA "704 Diamond" identification system.

NFPA 49, *Hazardous Chemicals Data*. Contains data on the fire hazards of several hundred chemical substances including suggested hazard identification numbers for use with the NFPA "704 Diamond" System of Identification.

NFPA 704, *Standard System for the Identification of the Fire Hazards of Materials*. Describes in detail the NFPA "704 Diamond" system for identifying health, fire, reactivity and other related hazards as might be encountered under fire or related emergency conditions.

Chapter 15

STORAGE AND HANDLING OF FLAMMABLE AND COMBUSTIBLE LIQUIDS

HAZARDOUS PROPERTIES

Achieving safe storage and handling of the great variety of flammable and combustible liquids commonly available requires that you familiarize yourself with some physical and chemical characteristics that affect or determine the relative degree of hazard associated with the liquid. The distinction between a flammable and a combustible liquid is somewhat arbitrary, and is based on flash point. It is important to remember that it is the vapor given off that burns, and not the liquid itself.

Flash Point

The flash point is the minimum temperature at which a liquid gives off vapor in sufficient concentration to form an ignitible mixture with air near the surface of the liquid. It is determined in a test apparatus as specified by an appropriate test procedure. The flash point, which indicates the tendency of a product to generate vapor, becomes a primary factor in determining fire hazard. Liquids with flash points below ambient storage temperatures usually have a rapid rate of flame spread, since it is not necessary for the heat of the fire to expend its energy in heating the liquid to form additional vapors.

Liquids with higher flash points are a lesser hazard because of a decreased chance of ignition and a decreased potential for vapor spread.

The flash point of a substance is generally a few degrees below its "fire point," since at the flash point temperature the vapors are not being generated fast enough to sustain combustion. As the term "flash point" suggests, the vapors generated at that temperature will flash but will not continue to burn. Fire point is not of major importance in considering hazardous characteristics of a liquid, as it is so close to flash point that, under operating conditions, such a precise difference is immaterial.

Classification of Flammable and Combustible Liquids

For classification purposes, distinction is made between a liquid and a gas, and also between a liquid and a solid. A liquid has a vapor pressure not greater than 40 psi absolute at 100°F. Substances with higher vapor pressures at that temperature are treated as gases. Liquids also have a specified fluidity, and substances with less fluidity are treated as solids. That measured fluidity is called 300 penetration asphalt, and the test involves placing a pointed weight of specified dimension on a sample of the material that has been warmed to 100°F. If the weight disappears in 300 seconds, the material is a liquid.

Flammable liquids have a flash point below 100°F, and combustible liquids have a flash point at 100°F or above. Further sub-classifications are made, and you should be aware of them.

Table 15-1.   Classification of Flammable Liquids

Classifi-cation	Flash Point (°F)	Boiling Point (°F)
IA	Below 73°	Below 100°
IB	Below 73°	100° and above
IC	73° to 100°	
II	100° to 140°	
IIIA	140° to 200°	
IIIB	200° and above	

Ignition Temperature

The ignition temperature is that to which a substance must be raised to ignite by itself. The test involves heating a closed or nearly closed container to the temperature at which a liquid will ignite spontaneously when introduced to the container. The ignition temperature of a liquid is generally several hundred degrees above its flash point temperature.

Temperature and Pressure Effects

Liquids are only slightly compressible and they cannot expand indefinitely. Liquids tend to become gases as the temperature increases, or as the pressure decreases.

Flammable Limits

The lower flammable limit is the minimum concentration of vapor to air below which propagation of a flame will not occur in the presence of an ignition source. The upper flammable limit is the maximum vapor to air concentration above which propagation of flame will not occur. If a vapor to air mixture is below the lower flammable limit, it is "too lean," and if it is above the upper flammable limit it is "too rich" to transmit flame. When the mixture is between the lower and upper limits, ignition can occur and explosion may result. In the mid-range between upper and lower limits, the ignition is generally more intense.

Rate of Evaporation

This is the rate at which a liquid vaporizes, at any given temperature and pressure. Differing rates of evaporation are of significance in fire protection. Generally, lower boiling points mean higher rates of evaporation and higher vapor pressure.

Vapor Density

Vapors from flammable and combustible liquids, in the pure state, are heavier than air. However, vapors in the pure state can only exist at or above the boiling point of the liquid. For all other conditions, the vapor is mixed with some air and the density is thereby proportionately reduced. Flammable vapors are likely to accumulate in basements and pits, unless there is means provided to drain off at the bottom. Storage of flammable liquids in basements should be avoided.

The vapor density of a liquid shows its relation to the weight of air. Thus, a vapor density of 2 indicates the pure vapor is twice as heavy as air. The higher the vapor density, the greater will be the tendency of the vapor to settle in low areas or to flow downhill. Large releases of high density vapor can flow substantial distances. Ignition at a distance will often result in a flashback to the vapor source.

Density and Water Solubility

Most flammable liquids are lighter than water and, if not miscible with water, will float on top. Carbon disulfide is an example of a liquid that is heavier than water and, therefore, water will float above it. Some flammable liquids, such as acetone and alcohol, will dissolve in, and completely mix with, water.

Boiling Point

The boiling point is the temperature of the liquid at which its vapor pressure equals atmospheric pressure. Above this temperature, the pressure of the atmosphere can no longer hold the liquid in a liquid state. The lower the boiling point of a liquid, the greater the vapor pressure and, consequently, the greater the fire hazard.

Viscosity

The viscosity of a liquid is a measure of its resistance to flow.

Identification

Identification of the hazard class of liquids is difficult. The sense of smell is not reliable. Many liquids are sold under names that give no indication of the potential fire hazard. Where doubt exists, a laboratory test for flash point will provide a positive determination. You can make a cursory estimate of hazard by placing a small sample in the bottom of a cup, removing it to a safe outdoor location, and trying to light the vapor in the cup. A flash of flame would indicate a low flash point of liquid. However, absence of a flame should not be accepted as proof that the liquid is safe.

Labeling of containers of flammable liquids is required in many jurisdictions. Absence of a label, however, is no proof that the liquid is not flammable. Where the Federal Hazardous Substances Act applies, it requires labeling based on open cup flash points as follows:

"Extremely flammable" indicates a flash point of 20°F or below;

"Flammable" means above 20°F to 80°F inclusive; and "Combustible" means above 80°F to 150°F inclusive.

This Act applies to consumer products.

STORAGE

The chief hazard connected with flammable liquid storage is accidental release of product. Such release usually results from container failure, and the failure may be caused in turn by overpressure, mishandling, corrosion, or puncture. Once accidental discharge of a flammable liquid occurs, the likelihood of ignition increases significantly. If the discharge takes place during a fire, or as the result thereof, then the severity of the incident will be magnified many times over.

Buried Tanks

Underground tanks with all connections through the top offer a very safe storage setting. Tanks can be buried inside buildings provided fill, vent, and gaging connections lead to the exterior of the building. Where corrosion poses a problem, the tanks should be either suitably protected or of corrosion resistant materials of construction. The excavation should not jeopardize the foundation of structures, or be placed where the settlement of the structure might damage the tank. Tanks should also be protected to prevent being dislodged where high ground water or flood water might be a problem.

Outside Aboveground Tanks

Outside aboveground tanks are suitable for large quantity storage. Some essential features include proper construction, well-engineered foundation, adequate emergency relief venting, and substantial pipe connections with an adequate number of properly placed valves to control flow in the event of fire or of breakage in the piping. Where natural drainage to a safe impounding area is not available, then the control of spillage can be achieved by diking. Dikes should be adequately sized and maintained in good condition. Openings for rain water drainage should normally be kept closed. Aboveground tanks should be properly spaced, both in relation to adjacent tanks and to nearby buildings and property. The greater the spacing, the less is the hazard.

Tanks Inside Buildings

Inside tank storage of flammable liquids is not recommended, but it is permitted in certain occupancies where such storage is necessary to the process. Tanks with weak roof seams are not allowed inside buildings, and floating roof tanks are similarly unacceptable. Emergency vents, where re-

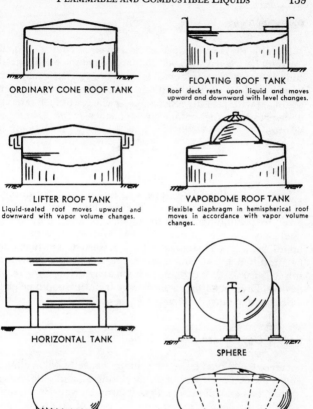

ORDINARY CONE ROOF TANK

FLOATING ROOF TANK
Roof deck rests upon liquid and moves upward and downward with level changes.

LIFTER ROOF TANK
Liquid-sealed roof moves upward and downward with vapor volume changes.

VAPORDOME ROOF TANK
Flexible diaphragm in hemispherical roof moves in accordance with vapor volume changes.

HORIZONTAL TANK

SPHERE

SPHEROID

NODED SPHEROID

Figure 15-1. Common types of tanks for the storage of flammable and combustible liquids.

quired, must terminate outside the building. Inside storage of fuel oil in tanks of not more than 660 gallons capacity, with fill and vent connections outside, is accepted practice.

Portable Tanks

Portable tanks are defined as closed vessels having a liquid capacity between 60 and 660 gallons and not intended for fixed installation. They are considered to be more desirable than 55-gallon drums for shipping and storage because they are equipped with pressure relief devices. The use of such tanks is widespread in the automotive, chemical, and paint industries.

Drums

Drums, usually of 60-gallon capacity, are used for storage of various flammable solvents. They are best stored outdoors away from buildings, or in small detached storage buildings used only for such storage. Where drum storage is inside buildings, a special storage room is preferred, and it should be protected by an automatic extinguishing system. Most storage drums do not come equipped with pressure relief devices. Drainage from drum storage facilities should be provided, and arranged to facilitate fire control.

Safety Containers

Safety containers have a maximum capacity of 5 gallons, and come equipped with a spring closing lid and spout cover so that the can will safely relieve internal pressure when subjected to fire exposure. This is the type of container that should be used for storing and dispensing small quantities of flammable liquids inside buildings. The safety can is not intended for use in settings where the periodic release of flammable vapors may create a hazardous atmosphere.

TRANSFER AND DISPENSING

Flammable and combustible liquids may be transferred by

Figure 15-2. Typical safety cans having pouring outlets with tight fitting caps or valves normally closed by springs.

pumps, gravity flow, hydraulic or compressed gas displacement, or by pumping systems. The latter is most often used for transferring large amounts of liquid. A closed pipe pumping system is the safest method of handling large quantity transferral.

Pumping Systems

Positive displacement pumps offer a tight shutoff, and they prevent siphoning of the liquid when not in use. Centrifugal pumps do not provide a tight shutoff when taking suction under head, and siphoning of the liquid may result.

Gravity Systems

Gravity transfer is often used where very volatile liquids that may cause vapor lock in pumping systems are being transferred. Where there is a large supply source, transfer by gravity should be avoided, since the hazard of continued flow is present. Where faucets are used on drums they should be of the spring-closing type, and such valves must never be blocked open.

Hydraulic Systems

Hydraulic systems employ water pressure to transfer liquid from the container. Such systems cannot be used where the product being transferred is miscible with water. In order to use this type of transfer system, the containers must be pressure vessels, and provision must be made to prevent over-pressurization.

Compressed Gas Displacement Systems

Compressed air must not be used since it increases the probability of a vapor-air explosion. Compressed gas displacement maintains the system under constant pressure, so a pipe failure or careless valve operation can result in spillage of a considerable amount of product.

Dispensing Systems

These generally involve the transfer of liquid from fixed piping systems, drums, or 5-gallon cans into smaller end-use containers. Release of some vapor is practically unavoidable. Therefore, it is best that such transfer take place in a designated area that is adequately protected and ventilated. Where dispensing must take place outside the designated area, only approved safety cans should be used.

LOSS CONTROL

The following basic loss control guidelines apply in principle to all operations where flammable and combustible liquids are stored and handled.

Confinement of Liquids

The major objective of an effective loss control program is confinement of liquids and vapors inside the containers. A second objective is to minimize the effects of an accidental release. The following steps will help to achieve these goals:

• Use equipment that is designed for flammable and combustible liquid storage. Such equipment should be vapor-tight, should have the minimum number of openings necessary, and should be designed to relieve excess internal pressure to a safe location.

• Equip open vessels and vessels with loose fitting covers with overflow drains and emergency bottom dump drains that are piped to a safe location.

• Handle small amounts of liquids in approved containers.

• Provide adequate drainage systems to prevent the flow of liquids into adjacent work areas.

Ventilation

Ventilation is a loss prevention measure that can prevent flammable liquid fires under normal operating conditions. Ventilation cannot, however, prevent ignitions where abnormal vapor releases occur.

The ventilation that is required for health safety for personnel greatly exceeds that required for firesafety.

Where ventilation is required for firesafety, it should be at the rate of 1 cfm per square foot of floor area and so designed as to provide a sweeping action across the entire floor area. It should be exhausted to a safe location. The system should be interlocked so that the operation using flammable liquids will shut down if the ventilation is inadequate.

Spot ventilation at the work site is also acceptable.

Control of Ignition Sources

All ignition sources should be controlled or eliminated in areas where flammable vapors may be present. Specially classified electrical equipment may be needed in some areas. (See Chapter 8.) Sources of ignition also include open flames, heated surfaces, smoking, cutting and welding, frictional heat, static sparks, and radiant heat.

Because of the ease of ignition of flammable vapors in the proper mix with air and the multiplicity of ignition sources, best precautions consist in limiting the evolution of flammable vapors and their dispersion into the atmosphere. Naturally smoking, open flames, cutting and welding, and hot work in general should be controlled whether or not there are flammable vapors present, since things other than flammable vapors can also be accidentally ignited.

Fire Protection

A wet-pipe automatic sprinkler system is the preferred basic fire control system for areas where flammable and combustible liquids are stored or handled. Large areas of containerized storage may require special sprinkler installations, such as sprinklers at each rack level as well as at the ceiling. Storage tanks, vessels, and process equipment may need deluge water spray systems.

All tank foundations and supports should be of fire resistive or protected steel construction.

In small confined areas or inside special equipment or vessels it may be desirable to provide special extinguishing systems to supplement the automatic sprinkler systems.

Appropriate portable fire extinguishers are necessary in the event of small liquid fires or fires in other combustibles.

Hydrants and small fire hose with adjustable stream nozzles should be provided in areas where flammable and combustible liquids are stored, handled, or used. Hose streams can be used to cool adjacent tanks and structures, to extinguish fires, and to wash down spills.

INSPECTION CHECK LIST FOR DRY CLEANING PLANTS

Type of solvent, flash point not lower than specified for class of equipment used.

Solvent storage in underground tanks or otherwise safely arranged.

Tanks, pipe and equipment tight and free from leaks.

Careful handling of solvent, no spills.

Tight covers on extractors and washers.

Safety cans used for spotting operations, minimum exposure of flammable liquids.

Static ground wires, where required, not loose or broken.

Emergency exits from dry cleaning rooms.

Provision of automatic extinguishing systems on dry cleaning units, dry tumblers when required by the solvent used.

Portable fire extinguishers, suitability, number, condition.

Condition of automatic sprinklers or other general extinguishing system provided.

Electrical equipment in good condition; types for hazardous locations where required.

Boilers or other heating equipment in separate cutoff area out of path of vapor travel.

Observance of "no smoking" rules and elimination of any other ignition source. Are contents of garments checked, for example, for matches?

SI UNITS

The following conversion factors are given as a convenience in converting to SI units the English units used in this chapter.

$$\text{\%}(°F - 32) = °C$$
$$1 \text{ gal} = 3.785 \text{ litres}$$
$$1 \text{ cfm} = 0.283 \text{ m}^3/\text{min}$$

BIBLIOGRAPHY

Henry, Martin F. ed., *Flammable and Combustible Liquids Code Handbook*, 1st ed., National Fire Protection Association, Quincy, MA, 1981. An explanation of the provisions of the *Flammable and Combustible Liquids Code* accompanied by the complete text of the code.

McKinnon, G. P. ed., *Fire Protection Handbook*, 15th ed., National Fire Protection Association, Quincy, MA, 1981. Section 4, Chapter 4 discusses the hazards, properties, and characteristics of flammable liquids and fire prevention methods.

McKinnon, G. P. ed., *Industrial Fire Hazards Handbook*, 1st ed., National Fire Protection Association, Quincy, MA, 1979. The fire hazards of flammable and combustible liquids are discussed in Chapter 39. Their hazards in relation to specific industrial processes are discussed in chapters dealing with the processes themselves.

NFPA Codes, Standards, and Recommended Practices. (See the latest *NFPA Codes and Standards Catalog* for the availability of current editions of the following documents.)

NFPA 30, *Flammable and Combustible Liquids Code*. Requirements for tank storage, piping, valves and fittings, container storage, industrial plants, bulk plants, service stations, and processing plants.

NFPA 32, *Standard for Dry Cleaning Plants*. A standard specifying safeguards incident to dry cleaning and dry dyeing.

Chapter 16

GAS HAZARDS

Because a gas is the only one of three states of matter that has no shape or volume of its own, it must be tightly confined in a container or pipe up until the moment it is to be used. By its nature, a gas is always striving to get out of confinement to reach its "natural" state. Its hazards, therefore, can be considered in terms of the hazards of the gas if it is released from confinement in a way other than intended. Control of the hazards, while somewhat amenable to treatment of gas after it is released, is far more effective if aimed at keeping the gas confined.

The hazard of released flammable gases is evident as a fire or combustion explosion resulting from ignition of a flammable gas-air mixture in a confined space. However, an accumulation of a nonflammable gas can cause asphyxiation by displacing air from a room. Some gases are chemically reactive, within themselves or upon contact with other materials, and can generate heat and reaction products that are hazardous. Others are physiologically reactive and are toxic or poisonous.

Because a gas is so light in weight, it takes a lot of it to do a certain job. Except for gas distributed through pipelines from supplier to consumer, gases must be stored in containers. In order to have storage containers of a size that is practical to move or be located in a reasonable size space, the gas must be concentrated. This is done by pressurizing (compressing) it or by converting it to the liquid state. In the liquid state, the volume of gas realized by converting it back to gas can be 200 to 1,000 times its volume as a liquid.

In the case of all compressed gases and for all liquefied gases with the exception of some very large (multi-million gallon) containers, the gas in a container at normal room temperatures is under pressure ranging from about 100 psi to 2,000 or 3,000 psi. Such a gas container represents an explosion hazard simply because of these pressures and independent of the chemical or physiological hazards of the gas inside.

In summary, then, it should be evident that control of gas hazards is dominated by measures designed to keep gas confined — including measures aimed at maintaining the integrity of containers by preventing their explosive failure.

GAS CONTAINER SAFEGUARDS

Gas containers are pipes, portable cylinders constructed to U.S. Department of Transportation (DOT) or Canadian Transport Commission (CTC) regulations, containers which are part of cargo vehicles or railroad tank cars (also constructed to DOT or CTC regulations), or pressure vessels constructed to the ASME Boiler and Pressure Vessel Code. Large, low pressure liquefied natural gas (LNG) containers are constructed to other standards, but these are of limited number and will not be addressed here.

DOT/CTC and ASME Code containers are built to working pressures reflecting the pressure of the gas contained. In the case of compressed gases (not containing gas in liquid form), the working pressure is as high as it can be considering the economics of container cost and container weight considerations — usually 2,000 to 3,000 psi. In the case of liquefied gases, the working pressure is determined by the vapor pressure of the liquid at liquid temperatures representative of ambient temperature extremes in the USA and Canada — usually -40° to 130°F.

In both compressed gas and liquefied gas containers, the actual pressure varies directly with the gas or liquid temperature, increasing and decreasing with increases or decreases in temperature. As temperatures exceed 130°F, so pressure can exceed the working pressure of the container. While safety factors can tolerate this to a degree, all gas containers are equipped with overpressure limiting devices to limit the pressure. The operation of these devices, of course, will result in release of gas and consequent hazard thereof.

If the source of heat is a fire exposure, the functioning of overpressure safety relief devices may not relieve the pressure sufficiently to prevent explosive container failure due to weakening of the container metal (usually steel or aluminum) by heating. This is particularly true of liquefied gas containers because pressure is likely to be left in the container when the metal loses its strength even though some will be lost through the safety device. This type of explosion is known as a BLEVE (Boiling Liquid-Expanding Vapor-Explosion).

The inspector, then, should look for any condition that is conducive to container failure from fire or other causes. These include:

• Location of combustibles in the vicinity which could produce fire exposure. These should be kept clear of containers containing *any* gas.

• Because release of flammable gases can lead to fire exposure, flammable gas containers should not be stored close to containers of other gases. As a general rule a 20-foot distance is needed to isolate a torch from a relief device. If this is not feasible, separation by a noncombustible barrier as high as the relief devices (usually about 5 feet) and having a fire resistance rating of at least ½ hour is a reasonable alternative.

• Containers should be in good condition free of dents, gouges, and signs of corrosion. Portable DOT/CTC containers are required to be requalified for service at intervals prescribed in the regulations — usually five to twelve years after manufacture. As each container must be marked with the date of manufacture and with the retest date, the inspector can check this readily. Containers not properly qualified should be removed from service.

• Check the overpressure safety relief device openings for signs of corrosion, paint accumulation or other debris. Insects love to build nests in relief device openings.

• Portable DOT/CTC containers have either caps or metal collars designed to protect the container valve from being broken off or damaged. Such a container should have this protection in place at all times except when it is hooked up in service.

• Compressed gas containers are usually of small

diameters compared with their lengths. Thus, they are rather easy to tip over. They should be secured when being transported and in storage. Liquefied gas containers are more stable but still should be secured when being transported.

Gas piping is fabricated and installed to NFPA standards for specific gases and applications and to the Code for Pressure Piping of the ASME. With the exception of natural gas and, to a lesser extent, LP-Gas, most gas piping is installed aboveground. The inspector should be alert to signs of corrosion, poor support, or locations where piping is subject to physical damage (especially where over-the-road or off-road vehicle movements occur).

RELEASED GAS SAFEGUARDS

By far the most common use for gases is as a fuel in gas burning appliances or industrial heating equipment (with air) or in cutting and welding processes (with oxygen). Another common use is for medical purposes (principally oxygen and nitrous oxide).

All of these uses, of course, involve the release of gas under controlled conditions. The inspector should be aware of the features of the utilization devices used to maintain control. These include flame-failure gas shutoff devices on gas appliances and industrial heating equipment and check valves, flame arrestors, and safety relief devices in oxygen-fuel gas systems (designed to prevent mixing of fuel gas and oxygen in piping and containers).

Nearly all domestic and commercial gas appliances are covered by ANSI or UL standards and tested and listed by the American Gas Association Laboratories or UL. The inspector should be wary if he finds one that is not so listed. If he finds one, the standard applicable to the type of appliance involved is very useful in determining its safety.

Larger and more specialized industrial equipment is not as well standardized and may be custom-made. There are NFPA standards for industrial ovens, furnaces and large boilers. FM and IRI have standards for other equipment.

Release of gas from containers and piping occurs, of course. A fundamental safeguard is to stop this escape. This is true whether or not the gas is burning, because a gas fire should be extinguished by stopping the leak or cutting off the gas flow (extinguishment by the use of extinguishers will result in unburned gas still being released).

All containers, of functional necessity, are fitted with a valve which can be closed manually. In some instances, this valve may have an automatic shutoff capability governed by excessive heat (fire) or flow rate (excess flow check valves). The inspector should check the operability of such valves.

Operation of overpressure safety relief devices should be restricted to true emergency conditions. As a common cause of such gas release is excessive ambient temperatures not due to fire exposure, the inspector should be alert to locations where temperatures above 130°F can occur. These include locations near heat producing equipment or in poorly ventilated structures.

Storage areas should be ventilated to limit the concentration of released gas (any gas).

LP-GAS BULK STATION

Features of the station's management and operation that affect firesafety include obvious items such as fire extinguishers and "No Smoking" signs as well as the following special features.

Tank Car Unloading

The railroad spur track should be at least 50 feet from the main line, and the station should be surrounded by fencing. Wheel blocks should be provided under the car. Valves should be provided in the tank car and in the liquid line to stop the flow of product in case of hose failure.

Truck Unloading

The unloading point should have a concrete guard to prevent damage to the piping and hose racks. Valves are required in the truck and in the unloading line to stop the flow of product in case of hose failure. A shed should be provided to protect the operator from the weather.

Storage Tanks at Station

Tanks, marked to show that they are the proper type, should be mounted on substantial foundations or cradles and placed to minimize exposure of buildings and equipment. Gaging devices and relief valves should be provided. The tank outlet should be equipped with a valve to stop the flow in the event a line breaks.

Pump Installation

Separate pumps and piping are required for butane and propane to prevent mixing the two. The pumps, valves, and fittings must be designed for LP-Gas service, and piping should be well supported. There should be a bypass in the pump discharge line. The installation should also have a pressure gage and a pressure relief valve.

Truck Filling

The loading area should be paved and level. Wheel blocks should be provided for the truck, and there should be a remote control switch that allows the driver to shut down loading instantly. There should be a valve in the discharge line, and a hose vent and relief valve on the discharge hose line. There should also be a vapor return hose line.

Cylinder Filling

The loading platforms should be at truck height to lessen the chance that cylinders will be dropped. The loading platform should be built on solid fill and should be well ventilated at the floor level. Accurate weighing scales are required to prevent overfilling the cylinders. The fill line should be equipped with a bypass and a pressure gage. A remote control switch that provides for immediate shutdown on the pump is required.

NONFLAMMABLE MEDICAL GAS SYSTEMS

Oxygen and nitrous oxide are used extensively in hospitals, nursing homes, dental offices, medical professional buildings, and other medical facilities for anesthesia, analgesia, and therapy purposes. They will be stored and used in both portable apparatus and in permanent fixed piping systems, which pervade the facility extensively.

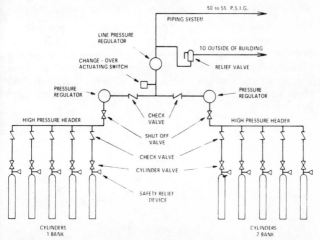

Figure 16-1. Typical cylinder supply system for nonflammable medical gas.

You will find great concern by facility owners and operators about the reliability of the supply and distribution of these gases because of their role in life safety. Because oxygen and nitrous oxide are nonflammable, their hazard as oxiding agents is sometimes not recognized. Both will lower the ignition temperature and accelerate combustion of flammable and combustible liquids, gases, and solids. Nitrous oxide, in addition, can decompose explosively in a container if subjected to fire temperatures, and it has an anesthetizing property if inhaled.

For both gases, the key firesafety precept is to keep combustibles away from them. Storage areas should be constructed of noncombustible or limited combustible materials, and there should be no combustible materials in the storage location — either in storage or in the fittings, e.g., shelves or racks.

Piping for such systems must be specifically cleaned to remove combustible matter, and joint brazing filler metal must have a melting point of at least 1,000 °F to minimize release of oxygen or nitrous oxide into a fire exposing the piping.

A certain quantity of these gases in the atmosphere is inevitable in the locations where these gases are administered to patients. The geometry of these locations is spelled out in standards, and ignition sources must be kept out of them. You should ascertain the administrative control procedures used in the facility.

SI UNITS

$$1 \text{ ft} = 0.305 \text{ m}$$
$$1 \text{ psi} = 6.895 \text{ kPa}$$
$$\tfrac{5}{9}(°F - 32) = °C$$

BIBLIOGRAPHY

American National Standards Institute. The following ANSI

documents deal with a wide variety of gas appliances and accessories.

ANSI Z21.1, *Household Cooking Gas Appliances*.

ANSI Z21.5.1, *Gas Clothes Dryers*, Volume I, Type 1 Clothes Dryers.

ANSI Z21.5.2, *Gas Clothes Dryers*, Volume II, Type 2 Clothes Dryers.

ANSI Z21.6, *Domestic Gas-Fired Incinerators*.

ANSI Z21.9, *Domestic Gas Hot Plates and Laundry Stoves*.

ANSI Z21.10.1, *Gas Water Heaters*, Volume I, Automatic Storage Type Water Heaters with Inputs of 75,000 Btu per Hour or Less.

ANSI Z21.10.3, *Gas Water Heaters*, Volume III, Circulating Tank, Instantaneous and Large Automatic Storage Type Water Heaters.

ANSI Z21.11.1, *Gas-Fired Room Heaters*, Volume I, Vented Room Heaters.

ANSI Z21.11.2, *Gas-Fired Room Heaters*, Volume II, Unvented Room Heaters.

ANSI Z21.13, *Gas-Fired Low-Pressure Steam and Hot Water Boilers*.

ANSI Z21.19, *Refrigerators Using Gas Fuel*.

ANSI Z21.40.1, *Gas-Fired Absorption Summer Air Conditioning Appliances*.

ANSI Z21.42, *Gas-Fired Illuminating Appliances*.

ANSI Z21.44, *Gas-Fired Gravity and Fan Type Direct Vent Wall Furnaces*.

ANSI Z21.47, *Gas-Fired Central Furnaces* (Except Direct Vent and Separated Combustion System Central Furnaces).

ANSI Z21.48, *Gas-Fired Gravity and Fan Type Floor Furnaces*.

ANSI Z21.49, *Gas-Fired Gravity and Fan Type Vented Wall Furnaces*.

ANSI Z21.50, *Vented Decorative Gas Appliances*.

ANSI Z21.55, *Gas-Fired Sauna Heaters*.

ANSI Z21.56, *Gas-Fired Swimming Pool Heaters*.

ANSI Z21.58, *Outdoor Cooking Gas Appliances*.

ANSI Z21.60, *Decorative Gas Appliances for Installation in Vented Fireplaces*.

ANSI Z21.61, *Gas-Fired Toilets*.

ANSI Z21.64, *Direct Vent Central Furnaces*.

ANSI Z21.65, *Separated Combustion System Central Furnaces*.

ANSI Z83.4, *Direct Gas-Fired Make-Up Air Heaters*.

ANSI Z83.6, *Gas-Fired Infrared Heaters*.

ANSI Z83.8, *Gas Unit Heaters*.

ANSI Z83.9, *Gas-Fired Duct Furnaces*.

ANSI Z83.10, *Separated Combustion System Central Furnaces*.

ANSI Z83.11 *Hotel and Restaurant Gas Ranges and Unit Broilers*.

ANSI Z83.12, *Commercial Gas Baking and Roasting Ovens*.

ANSI Z83.13, *Hotel and Restaurant Gas Deep Fat Fryers*.

ANSI Z83.14, *Gas Counter Appliances*.

ANSI Z83.15, *Gas-Fired Kettles, Steam Cookers and Steam Generators*.

ANSI Z21.2, *Gas Hose Connectors for Portable Indoor Gas-Fired Appliances*.

ANSI Z21.12, *Draft Hoods*.

ANSI Z21.15, *Manually Operated Gas Valves*.

ANSI Z21.17, *Domestic Gas Conversion Burners*.

ANSI Z21.18, *Gas Appliance Pressure Regulators*.

ANSI Z21.20, *Automatic Gas Ignition Systems and Components*.

ANSI Z21.21, *Automatic Valves for Gas Appliances*.

ANSI Z21.22, *Relief Valves and Automatic Gas Shutoff Devices for Hot Water Supply Systems*.

ANSI Z21.23, *Gas Appliance Thermostats*.

ANSI Z21.24, *Metal Connectors for Gas Appliances*.

ANSI Z21.35, *Gas Filters on Appliances*.

ANSI Z21.41, *Quick-Disconnect Devices for Use with Gas Fuel*.

ANSI Z21.45, *Flexible Connectors of Other Than All-Metal Construction for Gas Appliances*.

ANSI Z21.54, *Gas Hose Connectors for Portable Outdoor Gas-Fired Appliances*.

ANSI Z21.66, *Electrically Operated Automatic Vent Damper Devices for Use with Gas-Fired Appliances*.

ANSI Z21.67, *Mechanically Actuated Automatic Vent Damper Devices for Use with Gas-Fired Appliances*.

ANSI Z21.68, *Thermally Actuated Automatic Vent Damper Devices for Use with Gas-Fired Appliances*.

ANSI Z21.69, *Connectors for Movable Gas Appliances*.

McKinnon, G. P. ed., *Fire Protection Handbook*, 15th ed., National Fire Protection Association, Quincy, MA, 1981. Several sections of the Handbook contain data and fire protection information on fuel gases and their use in various industrial processes.

McKinnon, G. P. ed., *Industrial Fire Hazards Handbook*, 1st ed., National Fire Protection Association, Quincy, MA, 1979. Several chapters deal with industrial applications of fuel gases.

NFPA Codes, Standards, and Recommended Practices. (See the latest NFPA Codes and Standards Catalog for the availability of current editions of the following documents.)

NFPA 50, *Standard for Bulk Oxygen Systems at Consumer Sites.* Standard on location distance between bulk systems and exposure, containers and associated equipment.

NFPA 50A, *Standard for Gaseous Hydrogen Systems at Consumer Sites.* Covers containers, safety relief devices, piping, etc.

NFPA 50B, *Standard for Liquefied Hydrogren Systems at Consumer Sites.* Covers containers, supports, safety releases, etc.

NFPA 51, *Standard for the Design and Installation of Oxygen-Fuel Gas Systems for Welding and Cutting.* Addresses acetylene, hydrogen, natural gas, LP-Gas, MAP and other stable gases.

NFPA 51A, *Standard for Acetylene Cylinder Charging Plants.* Safety requirements for acetylene cylinder charging plants.

NFPA 51B, *Standard for Fire Prevention in Use of Cutting and Welding Processes.* Management of fire prevention and protection precautions.

NFPA 54, *National Fuel Gas Code.* Criteria for installation, operation and maintenance of gas piping, appliances, and equipment.

NFPA 56B, *Standard for Respiratory Therapy.* Use of nonflammable medical gases at atmospheric pressure.

NFPA 56F, *Standard for Nonflammable Medical Gas Systems.* Covers piped oxygen, nitrous oxide, compressed air, and other nonflammable medical gases.

NFPA 58, *Standard for the Storage and Handling of Liquefied Petroleum Gases.* General provisions for LP-Gas equipment and appliances.

NFPA 59, *Standard for the Storage and Handling of Liquefied Petroleum Gases at Utility Gas Plants.* Construction and operation of LP-Gas equipment at plants owned by gas utilities.

NFPA 59A, *Standard for the Production, Storage, and Handling of Liquefied Natural Gas (LNG).* Siting, design, constructruction, and fire protection for LNG facilities.

Chapter 17

COMBUSTIBLE DUSTS

Most finely divided combustible materials and some finely divided metals are subject to rapid combustion when dispersed in air, usually resulting in destructive explosions. The ease with which these dust/air mixtures ignite and their ability to propagate flame and generate damaging explosion pressures depend on a number of factors. This chapter will not address these factors in any detail, nor will it present data for individual dusts. For more in-depth knowledge, you are referred to the works of Palmer and Thorne listed at the end of this chapter. Similarly, the U.S. Bureau of Mines has published extensive data on many types of dusts.

For the purpose of this manual, a dust particle is a piece of material ranging from 1 to 150 microns in diameter. Basically, this means any material fine enough to pass through a 100 mesh sieve and not necessarily of regular shape. Particles larger than 100 mesh can be considered "powder" or "grit" and can generally be expected to pose no unusual hazards. For example, sawdust is primarily a collection of variously sized tiny chips of wood. Although some very fine particles are present, they do not contribute much to the combustibility of the mass, and the sawdust presents little more than a fire hazard. In contrast, the fine, floury dust generated by a sanding operation can explode quite destructively. The sanding operations in plywood and particle board plants have experienced some severe dust explosions.

The concept of a dust explosion is sometimes not well understood because an explosion can result from a number of different physical, mechanical, and chemical processes. In terms of the burning characteristics of dusts, the word "explosion" means "deflagration." A deflagration is a combustion reaction, just like any other burning process that is quite rapid.

The combustion wave (or flame front) moves through the fuel/air mixture at less than the velocity of sound; for dust explosions, this is typically from 1 to 10 meters per second. Although this is much faster than the flame spread in a traditional fire, it is much slower than the pressure wave generated by the production and thermal expansion of combustion gases. The pressure wave may be moving at speeds up to sonic velocity, 330 meters per second. The pressure wave is responsible for most of the damage in a dust explosion and has two characteristics of importance: maximum pressure and maximum rate of pressure rise.

THE TYPICAL DUST EXPLOSION

Dust explosions can be very destructive, the most vivid examples being those that occur in large grain elevators. In most cases, there are actually a series of explosions and the first is frequently quite small and innocuous. Although small in size and usually the result of a minor process malfunction, this first explosion is intense enough to dislodge any static dust on walls, ledges, and other surfaces in the immediate vicinity. Remember, the pressure wave is moving much faster than the flame front; it is knocking dust loose and mixing it with air, creating a much larger dust cloud just in time to be ignited by the following flame front. This secondary explosion is much larger and quite destructive; it can rupture process equipment and blow through masonry walls very easily. It has the capability to trigger even larger subsequent explosions.

FIRE HAZARDS OF DUSTS

Combustible dusts are also a fire hazard. Stationary deposits of dust provide an easy means for a fire to spread rapidly from its initial location. This will probably allow the fire to spread out and extend into adjacent areas faster than a sprinkler system can react. Overtaxing the system and its water supply is a distinct probability in such situations.

Another characteristic of dust is its ability to act as a thermal insulation. Thick deposits of dust on heat-producing equipment such as motors, shaft bearings, etc., retard the flow of heat. The equipment runs hotter. If the dust is organic, it will begin to carbonize, a chemical change that tends to lower the dust's ignition temperature and increase its insulating ability. Eventually, the dust ignites or the equipment fails, with possible ensuing fire.

FACTORS INFLUENCING IGNITION

Particle Size

The smaller the particle size of the dust, the easier it is to ignite. This is because the ratio of surface area to volume increases tremendously as particle size decreases. The particle is less able to absorb energy from an external source. Once ignited, it radiates energy to nearby particles more quickly and more efficiently. It is also true that a decrease in particle size will increase the rate of pressure rise during an explosion and will decrease the minimum explosive concentration, the ignition temperature, and the energy necessary for ignition.

As a practical matter, a range of particle sizes will be present, and the behavior of the dust will depend on the distribution of particles within the range.

Concentration

There is a minimum concentration of dust in air below which propagating ignition will not occur. As stated above, this minimum concentration decreases with decreasing particle size. Plotting explosion pressures and rates of pressure rise against dust cloud concentration shows that both parameters are at a minimum value at the minimum explosive concentration, rise to a peak value at a concentration slightly greater than stoichiometric, then decrease again as the concentration increases further. Because of experimental difficulties, maximum explosive concentrations have never been determined and may be relatively meaningless.

The energy of the ignition source, turbulence in the dust cloud, and uniformity of dispersion all have some effect on the minimum explosive concentration, at least in laboratory tests. In field situations, these influences are not important.

Moisture

The humidity of the air surrounding the dust particles has no effect on the course of a dust explosion. The moisture contained within the dust particles, however, does tend to increase ignition temperature, ignition energy, and minimum explosive concentration and tends to decrease the explosion severity. Again, these variations are most significant in the laboratory, although a moist dust will burn much less efficiently than a dry one.

Inert Material

The presence of an inert dust will reduce the ignitibility of a combustible dust. This is because the inert dust absorbs energy with no detrimental effect. The only practical application of this fact is the rock dusting of coal mines to prevent mine explosions. In all other cases, the inert dust would have to be present in concentrations too high to be tolerated by process requirements.

Inert gas can, however, be used to prevent dust explosions in process equipment. The inert gas system must not be reactive with the combustible dust. For example, some metals will burn in atmospheres of pure carbon dioxide or nitrogen. Consult NFPA 69, *Standard on Explosion Prevention Systems*, for details of inert gas systems.

Other Factors

As would be expected, a decrease in the partial pressure of

oxygen, either by vacuum or inert gas, will decrease the explosion hazard of a dust. In the production of metal powders, enough oxygen must be maintained in the system to permit controlled oxidation of the surface of the particles. If this is not done, the metal particles will oxidize so rapidly when they are exposed to the air that they will self-ignite.

The presence of a flammable gas in the dust/air mixture greatly increases the hazard. These so-called "hybrid mixtures" will produce explosions that are far more violent than would be expected. They will also ignite under conditions where both the gas and the dust are below the lower flammable limit and minimum explosive concentration, respectively. Situations involving hybrid mixtures require special safeguards, such as inert gas protection or explosion suppression systems.

FACTORS INFLUENCING EXPLOSION SEVERITY

Ignition Sources

Dust clouds and dust layers can be ignited by all of the usual ignition sources, including open flames, electrical arcs, frictional and mechanical sparks, and hot surfaces. Ignition by static electrical charge is usually ruled out. Dust clouds usually require 10 to 40 millijoules of energy for ignition to occur, compared to 0.20 to 1.0 millijoules for most flammable gases and vapors. This is generally above the energy potential of most static sparks. Of course, these same levels are well below the energy content of the common ignition sources mentioned.

Turbulence

The presence of turbulence in the dust cloud tends to increase the violence of the explosion. This is because the turbulence tends to sweep hot combustion gases away from the burning particles, allowing oxygen to move in. Also, the hot gases are then better able to preheat unburned dust particles. The net effect is to increase the burning rate of the cloud.

Maximum Explosion Pressure

The peak pressures reached by most dust explosions under test conditions exceed 50 psi. These peaks are subject to variation due to particle size distribution, concentration, and other variables. When you consider that typical construction will only withstand a few psi, it becomes evident that even the most inefficient dust explosions will do considerable damage. In fact, most dust explosions are far from optimal. Rarely is there uniform dispersion or particle size distribution; the damage that typically results is done at well below the explosion pressures that could have been developed.

Rate of Pressure Rise

The rate of pressure rise is roughly the ratio of the peak pressure to the time interval during which the pressure increased. It is the most important factor in determining the severity of a dust explosion. The size of explosion vents are largely determined by this rate of rise. For very high rates of rise, explosion vents may not be practical and other protection devices may be necessary.

Confinement

The gaseous combustion products of a dust explosion expand at a rate as high as sonic velocity. In so doing, they exert significant pressures on the surrounding enclosure. Unless the enclosure is strong enough to withstand the peak pressure developed, it will fail. For this reason process equipment and the buildings in which they are located must be protected by explosion vents or a type of explosion prevention system.

Duration

Another factor in explosion severity is the duration that the explosion pressure exists. Consider the wall of a dust collector in which an explosion occurs. The wall "sees" a steady increase in pressure to some peak value over a finite time span, assuming rupture does not occur. Then the pressure begins to subside. If the pressure is plotted against time, the area under the curve is the total impulse imparted to the wall by the explosion. It is the total impulse, rather than the peak pressure, that ultimately determines the damage. This partly explains why dust explosions tend to be more damaging than gas explosions, even though they build up pressure more slowly and usually do not peak as high.

WHAT YOU NEED TO KNOW ABOUT DUST

When you are confronted with a dust explosion hazard, find out as much as you can about the material itself and the process conditions. Using Pittsburgh seam coal as a standard, the U.S. Bureau of Mines has developed a means of comparing relative hazards of different dusts. The rating system is explained in the *Fire Protection Handbook* and in U.S. Bureau of Mines publications listed at the end of this chapter. Using the data in those reports, you should be able to easily determine the relative hazard.

However, there are two precautions that must be noted. First, the Bureau of Mines data was generated using the Hartmann Apparatus. More recent data is generated in spherical vessels of 20-litre or 1-cubic meter size. Data generated in these vessels tend to be more consistent and are more readily sealed. Thus, the Hartmann data, while still valuable as a relative indication of hazard, is not as useful for predicting the effects of dust explosion vent requirements. Second, the particle size distribution of the dust in question may be so unlike the test dust that comparison may be futile.

If you are faced with a dust previously untested or if you are a victim of the above two considerations, then have a sample of the dust tested. You will want to know maximum explosion pressure, maximum rate of pressure rise, minimum explosion concentration and the concentration yielding the highest value of maximum pressure. You may also wish to determine the minimum ignition energy and the cloud and layer ignition temperatures.

PREVENTION OF DUST EXPLOSIONS

Dust Control

Eliminating, or at least greatly reducing, the amount of airborne and static dust is the single most important means for preventing dust explosions. Dusts or materials that generate dust should be handled to the greatest extent possible in closed systems, either pneumatic or mechanical. In order to be effective, these systems must be dust-tight. Look for evidence of dust around seams and joints of ducts or pipes and around access panels. If dust is evident, it may indicate a faulty gasket or mechanical damage to the conduit itself. Make sure hatches on bins and tanks are closed securely. Bear in mind that the inside of these systems, whether handling process material in bulk or collecting fugitive dust, contains a

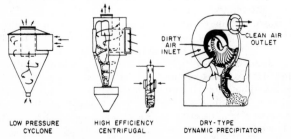

LOW PRESSURE CYCLONE

HIGH EFFICIENCY CENTRIFUGAL

DIRTY AIR INLET

CLEAN AIR OUTLET

DRY-TYPE DYNAMIC PRECIPITATOR

Figure 17-1. Examples of typical types of dust collecting equipment.

dust/air mixture in the explosive range; make sure there are adequate provisions for keeping tramp metal out.

Housekeeping is imperative. Any surfaces, horizontal or vertical, on which dust may accumulate should be kept clean. Frequency of cleaning will depend on conditions in the plant, such as the adequacy of dust collection systems and tightness of process equipment. Cleaning should be done by central vacuum systems. If portable units are used, they should be suitable for Class II, Division 2 areas. Alternatively, soft brushes and dust pans made from conductive plastic may be used. There are two rules of thumb for judging the adequacy of housekeeping. Look for dust deposits on a surface whose color contrasts sharply with the dust. If the color of the surface is totally obscured, cleaning needs to be done more often. If dust deposits exceed the thickness of a paper-clip, cleaning is long overdue.

Under no circumstances should dust be blown off with compressed air!

Equipment that handles or produces dusts should be as tight as possible. Where equipment is loaded or unloaded by dumping, local dust collection pickups should be installed. Dust control pickups should also be installed at drumming and bagging lines. Carefully review process schematics and the equipment itself to find these points where dust aspirators may be needed. Such locations include hatch covers on bins, tanks, or other vessels and frequently opened access panels.

Process Equipment

If a dust explosion is going to occur, chances are it will originate in a piece of equipment. It is impossible to prevent dust clouds from forming in equipment unless the material is handled as a slurry or a damp cakey solid. Here, maintenance is the key. Look for similar signs of abuse on drag or en-masse conveyors. Conveyor belts should show no signs of over-wear. Idler rollers should be free spinning to avoid frictional heating, and possible failure, of the conveyor belt.

Certain pieces of equipment are especially susceptible to dust explosions. These include mills, pulverizers, and the various types of dryers. Dust collectors and cyclone separators also fall into this category. One of the following forms of protection, or a combination, should be provided:

• The equipment can be designed to contain the expected explosion pressure;

• The equipment can be fitted with explosion vents;

• An explosion suppression system can be installed;

• The equipment can be purged with an inert gas.

Again, careful review of the process will help choose the most suitable option. In the case of the design-to-contain option, you may need the help of a qualified individual to determine if the equipment can withstand the expected overpressure. Explosion vents should never terminate within the building. Locate the equipment close to an outside wall and vent the equipment through a short, straight duct directly outside. Even better, for vented equipment, is to relocate outside or on the roof of the building. Breather vents can terminate within the building if fitted with filters. Check to make sure the filters are functioning properly.

Sources of Ignition

Assuming that a dust cloud will eventually form within the building, the next step is to prevent ignition. The most obvious two items are smoking and cutting and welding operations. Make sure all operational areas are posted for no smoking. Check to see that there are special areas set aside for safe smoking. There should be a hot work permit system as outlined in Chapter 24, Welding and Cutting. Ancillary equipment should be checked for overheating, slipping or chafing drive belts, or other mechanical defects that might cause sparks. Magnetic separators or screens may be needed if there is a possibility of tramp metal or other foreign objects entering equipment.

Electrical Equipment

Electrical equipment in dust-prone areas should be suitable

for Class II, Division 2 locations, unless the dust happens to be electrically conductive which requires *all* electrical equipment to be Division 1. Check to see that conduit is properly sealed, that all electrical enclosures are tightly sealed. Process control equipment that is located within the process stream must be suitable for Class II, Division 1 or must be intrinsically safe.

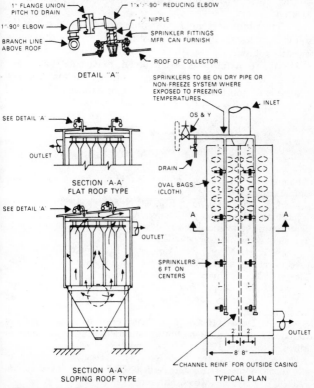

Figure 17-2. Sprinkler protection for a bag type dust collector.

FIRE FIGHTING

Other chapters in this Manual discuss what to look for when inspecting automatic sprinkler systems, hose connections, and portable fire extinguishers. In areas containing dust explosion hazards, the following must also be checked:

• Make sure all hose connections are equipped with fine spray nozzles. It may even be wise to specify the special fog nozzles used where high voltage electrical equipment is located.

• The plant personnel must understand that a coarse or solid water stream may throw dust into suspension, thus causing an explosion.

• Where metal dusts are involved, make sure that quantities of sand, talc, foundry flux, or other inert material are available to smother small fires. Make sure that water or portable extinguishers will *not* be used.

BIBLIOGRAPHY

Palmer, K. N., *Dust Explosions and Fire*, Chapman and Hall, Ltd., London, 1973.

Field, P., *Dust Explosions*, Elsevier, New York, 1982.

Jacobson, M., Nagy, J., Cooper, A. R., and Ball, F. J., "Explosibility of Agricultural Dusts," Report of Investigations 5753, U.S. Bureau of Mines, Pittsburgh, 1961.

Jacobson, M., Nagy, J., and Cooper, A. R., "Explosibility of Dusts Used in the Plastics Industry," Report of Investigations 5971, U.S. Bureau of Mines, Pittsburgh, 1962.

Jacobson, M., Cooper, A. R., and Nagy, J., "Explosibility of Metal Powders," Report of Investigations 6516, U. S. Bureau of Mines, Pittsburgh, 1964.

Nagy, J., Dorsett, H. G., Jr., and Cooper, A. R., "Explosibility of Carbonaceous Dusts," Report of Investigations 6597, U.S. Bureau of Mines, Pittsburgh, 1965.

Dorsett, H. G., Jr., and Nagy, J., "Dust Explosibility of Chemicals, Drugs, Dyes, and Pesticides," Report of Investigations 7132, U.S. Bureau of Mines, Pittsburgh.

Hertzberg, M., Cashdollar, K. L., and Opferman, J. J., "The

Flammability of Coal Dust-Air Mixtures," Report of Investigations 8360, U.S. Bureau of Mines, Pittsburgh, 1979.

Lathrop, J. K., "54 Killed in Two Grain Elevator Explosions," Fire Journal, Vol. 31, No. 4 (April 1971), pp. 17-29.

McKinnon, G. P. ed., Fire Protection Handbook, 15th ed., National Fire Protection Association, Quincy, MA, 1981, pp. 4-84 to 4-96.

Dorsett, H. G., Jr., Jacobson, M., Nagy, J., and Williams, R. P., "Laboratory Equipment and Test Procedures for Evaluating Explosibility of Dust," Report of Investigations 5624, U.S. Bureau of Mines, Pittsburgh, 1960.

Standard Test Method of Pressure and Rate of Pressure Rise for Dust Explosions in a Closed Vessel, ASTM E789-81, American Society for Testing and Materials, Philadelphia, 1981.

NFPA Codes, Standards, and Recommended Practices. (See the latest _NFPA Codes and Standards Catalog_ for the availability of current editions of the following documents.)

NFPA 61A, _Standard for Manufacturing and Handling Starch._ The hazards in handling dried starch and precautions to prevent ignition.

NFPA 61B, _Standard for the Prevention of Fire and Dust Explosions in Grain Elevators and Bulk Grain Handling Facilities._ A standard to prevent dust explosions and to minimize damage if an explosion should occur.

NFPA 61C, _Standard for the Prevention of Fire and Dust Explosions in Feed Mills._ Standard construction, ventilation, and equipment.

NFPA 61D, _Standard for the Prevention of Fire and Explosions in the Milling of Agricultural Commodities for Human Consumption._ The prevention of fire and dust explosion hazards in the milling of wheat, rye, barley, corn, etc.

NFPA 65, _Standard for the Processing and Finishing of Aluminum._ Covers operations in which fine aluminum dust or powder is liberated.

NFPA 66, _Standard for Pneumatic Conveying Systems for Handling Feed, Flour, Grain, and other Agricultural Dusts._ Pressure-type and suction-type systems.

NFPA 68, _Explosion Venting Guide._ Fundamentals of explosion venting, interpreting test data, vents and vent closures.

NFPA 69, _Standard for Explosion Prevention Systems._ Standard requirements, methods of application, inert gas supply and distribution.

NFPA 651, *Standard for the Manufacture of Aluminum or Magnesium Powder*. Covers flakes, powders, pastes, or atomized granules. Also covers the manufacture of explosive aluminum and magnesium alloys.

NFPA 654, *Standard for the Prevention of Dust Explosions in the Chemical, Dye, Pharmaceutical and Plastics Industries*. Reducing the hazards in the manufacture, fabricating, or molding of plastics.

NFPA 655, *Standard for the Prevention of Sulfur Fires and Explosions*. Crushing and pulverizing hazards in the handling of bulk and liquid sulfur.

NFPA 664, *Code for the Prevention of Dust Explosions in Woodworking and Wood Flour Manufacturing Plants*. A standard for the control of finely divided wood particles.

Chapter 18

METALS

Metals can burn. Some oxidize rapidly and reach flaming combustion; others oxidize so slowly that heat generated during oxidation dissipates before they can ignite. Certain metals, notably magnesium, titanium, sodium, potassium, calcium, lithium, hafnium, zirconium, zinc, thorium, uranium, and plutonium are called combustible metals. In thin sections, as fine particles, or molten metal they can ignite easily; however, in massive solid form they are comparatively difficult to ignite.

Aluminum, iron and steel are typical of metals that are not normally combustible. Yet they may ignite and burn in finely divided form. Clean, fine steel wool, for example, can be ignited. Dust clouds of most metals in air are explosive. Particle size, shape, quantity and alloy are important factors in assessing the combustibility of metals.

Temperatures in burning metals are generally much higher than in flammable liquids fires. Some hot metals can continue to burn in nitrogen, carbon dioxide or steam atmospheres, whereas fire in other materials would be extinguished.

All metals do not burn the same way. Titanium produces little smoke; smoke from lithium is dense and profuse. Some water-moistened metal powders, such as zirconium, burn with near explosive violence; yet the same powders wet with oil burn quietly. Sodium melts and flows; calcium does not. Some metals acquire an increased tendency to burn after prolonged exposure to moist air while exposure to dry air may make it more difficult to ignite the metal.

If you know the premises you are about to inspect has processes involving metals other than iron and steel, you should ascertain before the inspection the burning characteristics of the metals involved, the quantities involved in the processes, the processes themselves, the arrangement made for the storage and handling of scrap, and extinguishing agents used against combustible metal fires. A good guide is the NFPA

Fire Protection Handbook; and the NFPA standards governing the storage, handling, and processing of the various metals are the best sources for the additional information you will require for a good understanding of the intricacies involved in combatting hazards of metals (see the bibliography at the end of this chapter).

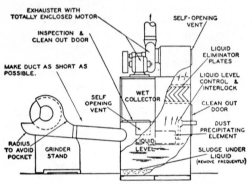

Figure 18-1. A schematic diagram of a water precipitation-type collector for use in safely collecting dry combustible metal dust.

EXTINGUISHING COMBUSTIBLE METAL FIRES

Combustible metal (Class D) fires are hard to extinguish and common extinguishing agents do not work well on them. Numerous agents are available to put out Class D fires, but a given agent does not necessarily work on all metals. Some agents work with several metals; others with only one. Commercially available agents are known as "dry powders" (not to be confused with "dry chemical" agents suitable for flammable liquid and live electrical equipment fires). Some powders go by their trade names, e.g., G-1 Powder, Met-L-X Powder, while others are materials known by their common names, e.g., talcum powder, sand, graphite, etc. Whatever the agent, its uses and limitations must be understood, and you should acquaint yourself with them through the references at the end of the chapter.

Controlling or putting out metal fires depends to a great degree upon the method of application and the training and experience of personnel who use them. In locations where combustible metals are present, ask to see the supply of dry powder agents on hand, their location and the tools available to spread them on burning metal. Inquire about the training given to employees in extinguishing metal fires. They involve techniques not commonly encountered in conventional fire fighting. Emphasize that training is needed to get experience in the specialized extinguishing techniques. Personnel with responsibility for controlling combustible metal fires should practice extinguishing fires in those metals at an isolated outdoor location.

MAGNESIUM

The ignition temperature of pure magnesium in large pieces is close to its melting point (1202 °F), but magnesium shavings, for example, can be ignited under certain conditions at about 950 °F, and finely divided powder at below 900 °F. Magnesium is principally used in alloy form, and certain magnesium alloys can ignite at temperatures as low as 800 °F. Thus, ignition temperatures can vary widely depending on the makeup of alloys involved. Find out as much as you can about the alloys used; it will help in sizing up the degree of hazard they represent.

Process Hazards

Magnesium and its alloys are readily machineable and, if the tools used are dull or deformed, frictional heat can ignite the chips and shavings created in the machining operations. If cutting fluids are used (machining magnesium is usually done dry), mineral oil types are the best. Water or water-oil emulsions are hazardous. Make sure water-based cooling liquids are not used by mistake and that the machines and surrounding work area are clean. Waste magnesium is best kept in clean covered, dry, steel or other noncombustible drums, which should be removed from the building at regular intervals. Magnesium fires have been known to take place in the bed of a machine.

Magnesium grinding is another hazardous operation, perhaps more so than machining. Magnesium dust clouds, made up of minute "fines" that could result from unprotected grinding operations, can be explosive if an ignition source is present. Make sure that grinding equipment used on magnesium has proper safeguards. An integral part of a good grinding installation is a water spray type dust precipitor that deposits the fines as a sludge in the collector housing. Fines from grinding generate hydrogen when submerged in water, but they cannot be ignited in this condition; however, grinding fines that are slightly wetted with water may generate enough heat to ignite spontaneously and burn violently as oxygen is extracted from the water with the release of hydrogen. Thus, it is important that grinding installations have interlocks that permit the grinder to operate only if the exhaust blower and water spray are working to keep the magnesium fines fully wetted.

Storage

Storage buildings housing magnesium preferably should be noncombustible and the magnesium segregated from combustible materials. Dry fines are stored in noncombustible containers in a fire resistive storage building or a room with explosion venting facilities. Check to make sure these arrangements are observed. Scrap magnesium fines wet with coolants should be stored outdoors in covered (but vented) noncombustible containers because of the possibility of spontaneous heating of the fines and hydrogen generation through reaction with the coolant.

Fire Extinguishment

The method of extinguishing magnesium fires depends mostly on the form of the material. Burning chips, shavings and small parts can be smothered and cooled with a suitable dry powder extinguishing agent. Fires in solid magnesium can be fought without difficulty if attacked in their early stages. Plenty of water, as from sprinkler discharge, on fire in solid magnesium cools the metal below the ignition temperature after a brief flare-up and the fire goes out rapidly; but if only a small amount is used, such as a fine water spray, the fire may intensify.

TITANIUM

Titanium, with ignition temperature ranging from 630°F in dust-air clouds to 2,900°F in solid castings, has hazard characteristics similar to magnesium. Castings and other massive pieces are not combustible under normal conditions; however, large pieces can ignite spontaneously in contact with liquid oxygen — truly an unusual condition not too often encountered. Small chips, fine turnings, and dust can ignite easily and burn with high heat release.

Process Hazards

Heat generated during machining, grinding, sawing and drilling of titanium may be enough to ignite the turnings and chips formed in the operations. Consequently, water-based coolants should be used in ample quantity to remove heat, and cutting tools should be left sharp. Fines should be removed regularly from the work area and stored in covered metal containers.

Grinding of titanium requires a dust collecting system discharging into a water-type dust collector. Large quantities of cooling fluid should be used, considerably more than considered normal, to keep down the sparking.

Descaling baths of mineral acids and molten alkali salts may cause violent reaction with titanium at abnormally high temperatures. Titanium sheets have also been known to ignite when they have been removed for descaling baths.

Storage

Large pieces need no special precautions. Dry scrap fines do require special precautions, such as storage in covered metal barrels kept well away from combustible materials. Moist scrap should be stored outdoors in covered metal barrels because of the possibility of hydrogen generation and spontaneous heating.

Fire Extinguishment

Good results on fires in drums and piles of titanium machining chips have been obtained with coarse water spray. The safest procedure in fires in small quantities of titanium powder is to ring the fire with one of the dry powders suitable for metal fires and to allow the fire to burn itself out. Again, good quantities of the right type of dry powder and a shovel or scoop to apply it should be handy.

ZIRCONIUM AND HAFNIUM

Massive pieces of zirconium do not ignite under ordinary conditions but will in the presence of a high oxygen concentration under certain conditions. Dust clouds of fires have been known to ignite at room temperature. Spontaneous heating and ignition is a possibility with scrap chips, borings, and turnings if fine dust is present.

Like other combustible metals, the combustibility of hafnium is related to its size and shape. Large pieces are hard to ignite; turnings and chips ignite easily. Hafnium is considered to be somewhat more reactive than titanium or zirconium. On ignition, hafnium burns with very little flame but with release of large quantities of heat.

Process Hazards

In general, processing recommendations for zirconium and hafnium are the same. To prevent dangerous heating during machining, large flows of mineral oil or water-based coolant are required. Turnings should be collected frequently and stored under water in cans.

Zirconium powder should be handled under an inert liquid or in an inert atmosphere. If either zirconium or hafnium powder is handled in air, extreme care must be used as the small static charges generated may cause ignition.

Storage

Zirconium castings do not require special storage precautions because massive pieces of the metal can withstand very high temperatures without igniting. Zirconium powder, however, is highly combustible, and it is customarily stored and shipped in one-gallon containers with at least 25 percent water by volume.

Storerooms for zirconium powder should be of fire resistive construction and equipped with explosion vents. Cans in storage should be separated from each other to limit fire spread. It's a good idea to check the cans periodically for corrosion.

Fire Extinguishment

Fighting fires in zirconium and hafnium require the same approach. Small quantities of the metals can be ringed with dry powder extinguishing agent and allowed to burn out. Fires in massive pieces of the metals can be fought with large quantities of water.

SODIUM, LITHIUM, NAK AND POTASSIUM

The principal fire hazard with sodium is its rapid reaction with water. Hydrogen liberated in the reaction may be ignited by the heat of the reaction itself. Once sodium is ignited it burns vigorously and forms dense clouds of caustic sodium oxide fumes.

Lithium undergoes many of the same reactions as sodium; however, its reaction with water is not so vigorous and not enough heat is generated to ignite the hydrogen given off in the reaction.

Fire hazards of potassium are very similar to those of sodium with the difference that potassium is usually more reactive.

NaK is the term used when referring to any of several sodium-potassium alloys. NaK alloys possess the same fire hazard properties as those of the component metals except that the reactions are more vigorous. All are liquids or melt near room temperature.

Process Hazards

A principal use of liquid sodium is as a heat transfer medium. Where molten sodium is used in process equipment, steel pans are located underneath to prevent contact and violent reactions of burning sodium with the moisture in concrete floors. Tray-type covers on the pans catch the sodium and drain it into the pans through drilled holes. Any sodium flowing through the holes extinguishes itself in the pan.

Information on sodium may be used as a guide in processing lithium, NaK, and potassium.

Storage

Sodium requires special precautions in storage because of its reactivity with water. Drums and cases preferably are stored in a dry, fire resistive room or building used exclusively for sodium storage. And, since sprinklers would be undesirable, no combustible material should be stored in the same area. Check to see that no water or steam pipes are located in the storage area and the sufficient heat is maintained to prevent moisture condensation. Natural ventilation at a high spot in the room can vent any hydrogen that may be released by accidental contact of sodium with moisture.

Storage recommendations for lithium, NaK and potassium in general parallel those for sodium.

Fire Extinguishment

Never use the common aqueous-based extinguishing agents on sodium fires — the reaction would be violent. The dry powders developed for metal fires, and dry sand, dry sodium chloride, and dry soda ash are effective. Sodium burning inside a piece of apparatus can usually be extinguished by closing all openings.

Fire extinguishing recommendations for sodium fires apply to fires in lithium, NaK, and potassium.

CALCIUM AND ZINC

The moisture in the air governs the flammability of calcium. If ignited in moist air it burns without flowing. Finely divided calcium will ignite spontaneously in air.

There are no serious fire hazards in sheets, castings or other massive forms of zinc because of the difficulty of ignition. Once ignited, though, zinc shapes can burn vigorously. Burning generates appreciable smoke.

METALS NOT NORMALLY COMBUSTIBLE

Aluminum

Because of its higher ignition temperature aluminum does

not have the same fire problems as magnesium. Only in powder or other finely divided form does it present a special fire problem, such as fires created in a grinding operation. Powdered or flaked aluminum under certain conditions can be explosive.

Iron and Steel

Iron and steel are not usually considered combustible; however, steel in the form of fine steel wool or dust may be ignited in the presence of excess heat, such as a torch. Ignition is more probable if the steel wool is saturated with a flammable solvent.

Fires have been reported in piles of steel turnings and other fine scrap which presumably contained some oil or other material that promoted self-heating. Spontaneous ignition of water-wetted barings and turnings in closed areas, such as ships' hulls, have been reported.

RADIOACTIVE METALS

Any metal can be made radioactive although some metals, e.g. uranium and thorium, are naturally radioactive. The important thing to remember is that radioactivity cannot be altered by fire and that radiation will continue wherever the radioactive metal may be spread during a fire. Smoke from fires involving radioactive materials frequently causes more property damage than the fire. The "damage" is radioactive contamination that must be cleaned up.

Uranium

Combustible, uranium is generally handled in such massive forms that it does not present a significant fire risk. In finely divided form it ignites easily and scrap from machining operations is subject to spontaneous heating. Grinding dust has been known to ignite, even under water, and fires have occurred spontaneously in drums of coarse scrap.

Thorium

Powdered thorium is usually compacted into small solid pellets and in that form can be safely stored or converted into alloys with other metals. Powdered thorium requires special handling techniques because of its low ignition temperature. The dry metal powder should not be in air because the friction of the particles falling through air or against the edge of a glass container may produce electrostatic ignition of the powder. Thorium powder is usually handled in a helium or argon atmosphere.

Plutonium

Plutonium is somewhat more susceptible to ignition than uranium. It is normally handled by remote control means and under an inert gas or "bone dry" air atmosphere. It is subject to spontaneous ignition in finely divided form, such as dusts and chips.

Plutonium metal is never intentionally exposed to water, in part because of fire considerations. Plutonium which ignites spontaneously is normally allowed to burn under conditions limiting both fire and radiological contamination spread.

BIBLIOGRAPHY

McKinnon, G.P. ed., *Fire Protection Handbook*, 15th ed, National Fire Protection Association, Quincy, MA, 1981. Section 4, Chapter 10, Metals, is a comprehensive resource on the fire hazards of metals; the safeguards to observe in their processing, storage, and handling; and fire extinguishing methods that can be used. Section 18, Chapter 5, Combustible Metal Agents and Application Techniques, describes the many different extinguishing agents that can be used on combustible metal (Class D) fires.

McKinnon, G.P. ed., *Industrial Fire Hazards Handbook*, 1st ed., National Fire Protection Association, Quincy, MA, 1979. Chapter 38, Metalworking, explains in detail the hazards involved in machining various metals and the safeguards that can be applied against fire hazards.

Purington, R.G. and Patterson, H.W., *Handling Radiation Emergencies*, National Fire Protection Association, Quincy, MA, 1977. A guide to handling fire emergencies involving radioactive materials in all forms.

NFPA Codes, Standards and Recommended Practices. (See the latest *NFPA Codes and Standards Catalog* for availability of current editions of the following documents.)

NFPA 48, *Standard for the Storage, Handling, and Processing of Magnesium*. Contains guidance on measures that can be taken to control the fire and explosion hazards in the storage, handling, and processing of magnesium and magnesium alloys.

NFPA 481, *Standard for the Production, Processing, Handling and Storage of Titanium*. Covers the properties and characteristics of titanium, sponge production, melting plants and mills, machine shops, powder production and use, special hazards, and extinguishment.

NFPA 482, *Guide for Fire and Explosion Prevention in Plants Producing and Handling Zirconium*. Contains information on the fire and explosion hazards of zirconium and the fire prevention and protection practices that can control the hazards. Emphasis is placed on dust collection and the disposal of zirconium scrap.

NFPA 65, *Code for the Processing and Finishing of Aluminum*. Covers operations where fine metallic aluminum dust or powder is liberated.

Chapter 19

CHEMICALS AND RADIOACTIVE MATERIALS

Safe and effective fire control measures when chemicals are involved require a knowledge of the hazardous properties of chemicals. For the purposes of this discussion, chemicals are classified according to their oxidizing abilities, combustibility, instability, reactivity with water or air, corrosiveness, and radioactivity. Although many chemicals possess more than one of these properties, it is customary to classify each by its dominant hazard.

OXIDIZING CHEMICALS

Most oxidizing chemicals are not combustible, but they may increase the ease of ignition of combustible materials and will increase the intensity of burning. Some oxidizing agents are susceptible to spontaneous decomposition.

Table 19-1. Typical Oxidizing Materials

Bromates	Nitrates
Chlorates	Perchlorates
Chlorites	Peroxides
Hypochlorites	Permanganates
Hydrosulfites	Persulfates

Nitrates

Inorganic nitrates are widely used in fertilizers, salt baths and other industrial applications. When exposed to fire, they may melt and release oxygen, causing the fire to intensify. Molten nitrates will react violently with organic materials. When solid streams of water are used for fire fighting, they may produce steam explosions upon contact with molten nitrates.

Nitric Acid

While nitric acid is generally considered a corrosive, it will markedly increase the ease of ignition of cellulosic materials when the acid or its vapors come in contact with such materials.

Nitrites

Nitrites are more active oxidizing agents than nitrates. Mixtures with combustible substances should not be subjected to heat or flame. Certain nitrites, notably ammonium nitrite, are by themselves explosive.

Inorganic Peroxides

Sodium, potassium, and strontium peroxides react vigorously with water and release oxygen and large amounts of heat. If organic or other oxidizable material is present when such a reaction takes place, fire is likely to occur.

Intimate mixtures of barium peroxide and combustible or readily oxidizable materials are explosive and easily ignited by friction or by contact with a small amount of water.

Hydrogen peroxide is a strong oxidizing agent and may cause ignition of combustible material with which it remains in contact. This is especially possible at concentrations above 35 percent. At a concentration above about 92 percent, hydrogen peroxide can be exploded by shock.

Chlorates

When heated, chlorates give up oxygen more readily than do nitrates. When mixed with combustible materials, they may ignite or explode spontaneously. Drums containing chlorates may explode when heated.

Chlorites

Sodium chlorite is a powerful oxidizing agent that forms explosive mixtures with combustible materials. In contact with strong acids, it releases explosive chlorine dioxide gas.

Dichromates

Among the dichromates, all of which are noncombustible, ammonium dichromate is the most hazardous. It begins to decompose at 356°F, and above 437°F decomposition becomes self-sustaining, releasing nitrogen gas. Closed containers rupture at the decomposition temperature.

Other dichromates release oxygen when heated and react readily with oxidizable materials.

Hypochlorites

Calcium hypochlorite may cause combustible organic materials to ignite on contact. When heated, it gives off oxygen. It is sold as bleaching powder or, when concentrated, as a swimming pool disinfectant.

Perchlorates

Mixtures of inorganic permanganates and combustible material are subject to ignition by friction, or they may ignite spontaneously in the presence of an inorganic acid.

Persulfates

These are strong oxidizing agents, which may cause explosions during a fire. An explosion may follow an accidental mixture of the persulfate with combustible material.

COMBUSTIBLE CHEMICALS

There are several chemicals that are hazardous principally because of their combustibility.

Carbon Black

Carbon black is most hazardous immediately following manufacture. After thorough cooling and airing, however, it will not ignite spontaneously, although it may generate heat in the presence of oxidizable oils.

Lamp Black

This material often ignites spontaneously when it is freshly bagged. It has a great affinity for liquids and heats in contact with drying oils. Lamp black should be stored in a cool, dry area away from oxidizing materials.

Lead Sulfocyanate

Lead sulfocyanate decomposes when heated. Its decomposition products include highly toxic and flammable carbon disulfide and highly toxic but nonflammable sulfur dioxide.

Nitroaniline

A combustible solid, nitroaniline melts at 295°F, and its flash point is 390°F. In the presence of moisture, it may cause the spontaneous ignition of organic materials.

Nitrochlorobenzene

This material, which is a solid at ordinary temperatures, gives off flammable vapors when heated.

Sulfides

Most sulfides are easily ignited. Antimony pentasulfide is hazardous in contact with oxidizing materials, while phosphorus pentasulfide may ignite spontaneously in the presence of moisture. Phosphorus sesquisulfide is considered highly flammable, having an ignition temperature of 212°F.

Sulfur

In liquid form, sulfur has a flash point of 405°F. Finely divided sulfur dust is an explosion hazard. Sulfur also forms highly explosive and easily detonated mixtures with chlorates and perchlorates.

Naphthalene

This chemical is combustible in both solid and liquid form. Its vapors and dust form explosive mixtures with air.

UNSTABLE CHEMICALS

Certain chemicals spontaneously polymerize, decompose, or otherwise react with themselves in the presence of a catalytic material, or even when pure. Such reactions may become violent.

Acetaldehyde

Acetaldehyde undergoes an addition-type reaction, which can become dangerous in the presence of certain catalysts and at elevated temperatures.

Ethylene Oxide

Ethylene oxide may polymerize violently when catalyzed

by anhydrous chlorides of iron, tin, or aluminium; oxides of iron (iron rust) and aluminum; and alkali metal hydroxides. It also reacts with alcohols, organic and inorganic acids, and ammonia.

Hydrogen Cyanide

This chemical is flammable and poisonous. In either the liquid or vapor state, it has a tendency to polymerize. The reaction is catalyzed by alkaline materials, and since one of the products of the reaction is alkaline, an explosive reaction will eventually take place.

Nitromethane

Nitromethane decomposes violently at 599 °F and 915 psig. It has been recognized that undiluted nitromethane may detonate under certain conditions of heat, pressure, shock, and contamination.

Organic Peroxides

Organic peroxides are combustible, and they increase fire intensity. Many organic peroxides can be decomposed by heat, shock, or friction. Some are detonatable.

Styrene

The polymerization reaction of styrene increases as temperature increases. Eventually the reaction will become violent, as it is accelerated by its own heat.

Vinyl Chloride

This is a flammable gas that may polymerize under fire conditions and cause violent rupture of its container.

WATER- AND AIR-REACTIVE CHEMICALS

The heat liberated during reactions of certain chemicals with air or water can be high enough to ignite the chemical if it is combustible. If the chemical is not combustible, the heat of reaction may ignite nearby combustible materials.

Alkalies

Caustics or alkalies, though noncombustible, will react with water and generate sufficient heat to ignite combustibles.

Aluminum Trialkyds

These metals for the most part are pyrophoric; that is, they ignite spontaneously on exposure to air, and react violently with water and certain other chemicals.

Anhydrides

Acid anhydrides are compounds of acids from which water has been removed. They react with water, usually violently to regenerate acids.

Carbides

Carbides of some metals may react explosively on contact with water. Many decompose in water to form acetylene.

Charcoal

Under some conditions, charcoal reacts with air at a rate sufficient to cause the charcoal to heat and ignite.

Hydrides

Most hydrides are compounds of hydrogen and metals. Metal hydrides react with water to form hydrogen gas.

Oxides

Oxides of metals can react with water to form alkalies and acids respectively.

Phosphorus

White phosphorus is more dangerous than red phosphorus because of its ready oxidation and spontaneous ignition in air. White phosphorus is very toxic and should not be permitted to come into contact with the skin.

Sodium Hydrosulfide

On contact with moisture and air, sodium hydrosulfide heats spontaneously and may ignite nearby combustible materials.

CORROSIVE MATERIALS

Corrosive materials are those which have a destructive effect on living tissues. Although they are usually strong oxidizing agents, they are separately classified to emphasize their injurious effect upon contact or inhalation.

Table 19-2.   Typical Corrosives

Acetic Acid	Hydrofluoric Acid
Anhydrous Ammonia	Hydrogen Peroxide (35-52%)
Antimony Pentachloride	Nitrating Acid
Benzoyl Chloride	Nitric Acid
Bromine	Oleum
Calcium Hypochlorite	Perchloric Acid (not over 72%)

Chlorine

Chlorine Trifluoride

Chromic Acid Solution

Fluorine

Hydrochloric Acid (muriatic acid)

Potassium Hydroxide (caustic potash)

Sodium Hydroxide (caustic soda)

Sulfur Chloride

Sulfur Trioxide

Sulfuric Acid

Inorganic Acids

Concentrated aqueous solutions of inorganic acids are not in themselves combustible. Their chief hazard lies in the danger of leakage and possible mixture with other chemicals or combustible material stored in the vicinity which would be followed in some cases by fire or explosions.

Halogens

Members of the halogen family — fluorene, chlorine, bromine, and iodine — differ from each other and decrease in the order named. The last two have the lesser fire hazards. Halogens are noncombustible, but will support combustion.

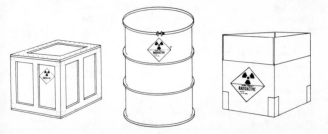

Figure 19-1. Three styles of shipping packages for radioactive materials.

FLAMMABLE SOLIDS

Some solid substances are likely to cause fire or explosion through friction, absorption of moisture, or exposure to air or moderate heat. Some such chemicals are listed below.

Danger due to **friction**:

Ammonium Perchlorate
Barium Chlorate
Benzoyl Peroxide (dry)
Potassium Chlorate
Potassium Perchlorate
Phosphorus (red, white, or yellow)

Phosphorus Pentasulfide
Sodium Chlorate
Sodium Perchlorate
Sodium Peroxide
Sulfur

Zinc Chlorate

Danger from heating due to **absorption of moisture**:

Aluminum Carbide
Aluminum Chloride
Aluminum Dust
Barium Peroxide
Bronze Dust
Calcium
Calcium Carbide
Calcium Oxide
Ferric Sulfide
Lithium
Lithium Hydride

Magnesium (finely divided)
Phosphorus Pentasulfide
Potassium
Potassium Peroxide
Selenium
Sodium
Sodium Hydride
Sodium Hydrosulfite
Sodium Peroxide
Zinc Dust

Dangerous in **air**:

Diborane
Lithium
Phosphorus (white or yellow)

Pyrophoric Iron
Sodium Hydride
Zirconium Dust

Dangerous from subjection to **moderate heat**:

Ammonium Perchlorate*
Antimony Pentasulfide
Barium Chlorate*
Calcium
Cellulose Nitrate (nitrocellulose)
Dinitroaniline
Dinitrobenzene

Phosphorus Pentasulfide
Phosphorus Sesquisulfide
Potassium Chlorate*
Potassium Perchlorate*
Sodium Chlorate*
Sodium Perchlorate*
Zirconium Dust

*when mixed with combustible or organic materials.

POISONS

Most chemicals are not considered to be toxic unless taken orally or inhaled in relatively large quantities. A few, however, can cause serious illness or death by external bodily contact, or if swallowed or inhaled in small quantities.

Gases and vapors that are dangerous to life where small amounts are mixed in air are considered to be extremely dangerous poisons. These include:

Acrolein	Dimethyl Sulfate
Bromoacetone	Fluorine
Carbon Disulfide	Hydrogen Cyanide
Chloropicrin	Methyl Bromide
Cyanogen	Nitrogen Tetroxide
Diborane	Phosgene

Less dangerous poisons that are hazardous when taken orally or on contact include:

Aldrin	Hydrogen Fluoride
	Methylamines
Allyl Alcohol	Mercury Compounds
Ammonia	Nicotine
Arsenic Compounds	Nicotine Sulfate
Beryllium	Nitrobenzene
Chlorine	Nitrochlorobenzene
Cresol	Parathion
Cyanides	Phenol (carbolic acid)
Dimethyl Sulfide	Procylamine
Dinitrobenzene	Strychnine
Dinitrochlorobenzene	Tetraethyl Lead
Endrin	Thallium Salts
Ethylamine	

RADIOACTIVE MATERIALS

Radioactive elements and compounds have fire and explosion hazards identical to those of the same materials when they are not radioactive. However, the presence of radiation may interfere with manual fire fighting efforts because of the threat it poses to humans. Salvage procedures and resumption of normal operations may be delayed because of the need to decontaminate buildings, equipment, and materials.

Radiological contamination can sift through openings and ventilation systems in the form of dust or vapor and spread throughout the structure. Persons who have been in a contaminated area should be isolated until they have been decontaminated. Fire fighters and other emergency personnel must be thoroughly trained and provided with full protective clothing including self-contained breathing apparatus.

Facilities handling radioactive materials or housing nuclear reactors or radiation machines should have a water supply that is adequate for fire control and decontamination operations. An automatic sprinkler system or specially designed water spray system is the first choice for fire protection, for it can operate in an area where the contamination level is so high that fire fighters cannot enter. Provision must be made for the control and collection of water run-off to avoid spreading contamination.

For the proper guidance on the handling of radioactive material, you should contact the director of your local emergency preparedness unit.

SI UNITS

The following conversion factor is given as a convenience in converting to an SI unit the English unit used in this chapter.

$$\frac{5}{9}\,(°F - 32) = °C$$

BIBLIOGRAPHY

McKinnon, G. P. ed., *Industrial Fire Hazards Handbook*, 1st ed., National Fire Protection Association, Quincy, MA, 1979. Chapter 37 discusses radioactive materials, while Chapter 4 deals with fire hazards in nuclear energy plants. Chapter 28 is devoted to the fire hazards of chemical processes.

McKinnon, G. P. ed., *Fire Protection Handbook*, 15th ed., National Fire Protection Association, Quincy, MA, 1981. Section 4, Chapter 6, deals with the fire and explosion hazards of chemicals.

NFPA Codes, Standards, and Recommended Practices. (See the latest *NFPA Codes and Standards Catalog* for the availability of current editions of the following documents.)

NFPA 43A, *Code for the Storage of Liquid and Solid Oxidizing Materials*. Establishes four classes of oxidizing materials and their storage requirements.

NFPA 43C, *Code for the Storage of Gaseous Oxidizing Materials*. Covers oxidizing materials that are gases at ambient temperatures or liquids having vapor pressures exceeding 40 psi at 100°F.

NFPA 49, *Hazardous Chemicals Data*. Defines hazards of chemicals, and discusses unusual storage or fire fighting problems. Data includes life and fire hazard precautions.

NFPA 490, *Code for the Storage of Ammonium Nitrate*. Suggested regulations for the storage of ammonium nitrate in bags, drums, or other containers or in bulk.

NFPA 491M, *Manual of Hazardous Chemical Reactions*. Contains information on over 3,500 potentially dangerous chemical reactions.

NFPA 495, *Code for the Manufacture, Transportation, Storage and Use of Explosives and Blasting Agents*. Provides regulations for safety in the manufacture, storage, transportation, and use of explosives and blasting agents commonly used in mining, quarrying, road building, harbor improvement and similar operations.

NFPA 655, *Standard for the Prevention of Sulfur Fires and Explosions*. A standard on the crushing and pulverizing hazards in handling bulk and liquid sulfur.

NFPA 704, *Standard for the Identification of the Fire Hazards of Materials*. A method of marking containers to indicate reactivity, flammability, and health hazards of materials.

NFPA 801, *Recommended Fire Protection Practice for Facilities Handling Radioactive Materials*. Practices aimed at reducing the risk of fire and explosion and minimizing the associated damage by radioactive contamination.

Chapter 20

PLASTICS

If you are responsible for inspecting a plant that processes, converts or fabricates plastics, you should be thoroughly informed about the characteristics of these materials and their susceptibility to fire. At present, there are about thirty major groupings or classes of plastics or polymers, but there is great variation in the composition of the thousands of individual products in the world markets. It is essential for you to know what materials are processed and stored in your plant, how to perform an inspection efficiently, and how to recommend corrective measures.

Plastics in solid form are considered to have burning characteristics similar to wood, but some produce more intense smoke than wood. As well as producing carbon monoxide, common to most burning materials, some plastics produce other toxic gases. The rate of release and quantity of these fire products can vary considerably, depending upon the composition of the individual plastic. Cellulose nitrate is the singular exception to the classification of plastics as an ordinary combustible, because its burning process requires large quantities of water for extinguishment.

Pelletized and liquid plastics also vary in burning characteristics and in their behavior when simply exposed to the heat of a fire or other high temperature. Plastic dust must also be considered a dangerous hazard because of its susceptibility to ignition and explosion. In your plant, the amount, kind, manufacturing, and storage of such products are important factors that determine the relative firesafety.

THE GENERAL SURVEY

Clean and neat housekeeping is important in any manufacturing plant because it is a necessary method of reducing the sources of fire or explosion. In a plastics-making plant, there can be accumulations of dusts and solid particles, flammable liquids, cloths, empty containers, and other combustibles that might contribute to a fire incident. These common hazards should receive your first attention, and the obvious correction is to have them removed and, if possible, alter the practices that cause their accumulation. Dust requires special removal, particularly if there is open flame or some other source of ignition. Improper sweeping or brushing might throw the dust into suspension so that it can ignite into flame or with explosive force.

The three general types of plastics manufacturing are synthesizing, conversion, and fabrication, and the procedures required for each can have some influence on the level of plant firesafety.

Synthesizing, or manufacturing, is the mixing of the basic plastic materials, or feedstock, sometimes with added coloring agents or other substances.

Conversion is the process of molding, extrusion, or casting the plastic so that it will flow into a certain shape that will be retained after cooling. Considerable heat is required for each of these processes.

Fabricating includes bending, machining, cementing, decorating, and polishing plastic, sometimes with the use of other materials that may be flammable, reactive, or perhaps fire retardant. For example, the thermoplastic ABS combines acrylonitrile, butadiene and styrene, two low flash point flammable liquids and a gas having an ignition temperature of 804°F. A molding process may combine a flammable resin with nonflammable glass fibers, or with heat resistant silicones.

Among the hazards to be considered in these operations are the combustible dusts, flammable solvents, electrical faults, hydraulic fluids, and the storage and handling of large quantities of combustible raw materials and finished products. As an inspector, you can refer to the standards and other publications of the NFPA, Underwriters Laboratories Inc., Factory Mutual Research Corporation, and many other reliable sources of technical information that will help you evaluate a given situation.

In the plastics manufacturing process, thermoplastic compounds are usually melted by heat and then forced into a mold or die for shaping. In original form, these compounds may be pellets, granules, flakes, or powder, and each of these forms can produce dust.

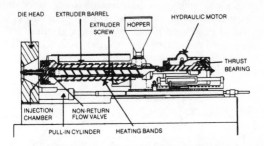

Figure 20-1. Diagram of a reciprocating screw injection molding machine in which plastic pellets are compacted, melted, and injected into a die, where the molten plastic is allowed to cool and harden. (Society of the Plastics Industry.)

Thermosetting resins may be in liquid form or as a partially polymerized molding compound, but for either form, considerable heat is needed for the molding process.

Other processes include: blow molding for the making of hollow products, such as bottles, gas tanks, and carboys; calendering, or the process of converting thermoplastics into film or sheeting, or applying a plastic coating to textiles or other materials; casting, using thermoplastics or thermosets to make products, by pouring the hot liquid solution into a mold, then letting it cool to solid form; coating, such as applying thermoplastic or thermosetting materials to metal, wood, paper, glass, fabric, or ceramics; compounding, or the mixing of additives with resins by kneading mixers or screw extruders; compression molding, or the use of heat and pressure to squeeze or press material into a certain shape; extrusion, or the movement of a thermoplastic material by screw thread or other propulsion to form a continuous sheeting, film, rod, cable, cord, or other product; foam plastics molding, or the use of foam plastics in casting, calendering, coating or rotational molding; high pressure laminating, using heat and pressure to join materials; injection molding, or the impinging of two or more high pressure reactive streams in a mixing chamber, then injecting the mixture into a mold; reinforced plastics processing, or combining resins with reinforcing materials; rotational molding, or the movement of powdered plastic or molding granules within a moving, heated container; and transfer molding, or the curing of thermosetting plastics in a mold under heat and pressure.

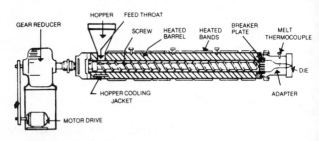

Figure 20-2. A basic single-screw extruder in which plastic pellets are fed from a hopper, driven forward, and melted. The molten plastic is fed through the adapter into the die. (Society of the Plastics Industry.)

As inspector, you would have to be concerned about the temperatures of these processes, the fire characteristics of the solids and liquids used, the type and condition of portable and automatic fire extinguishing equipment, and the availability of manual and automatic fire alarm equipment.

Your general survey should also include examination of aisles and corridors to see if they are blocked unnecessarily by equipment or vehicles or combustible products; notation of exit signs and facilities; examination of all fire extinguishers for pressure adequacy and dates of last inspection; and whether a fire escape plan is posted in one or more prominent locations of the work area. It may also be helpful to ask a few casual questions to learn whether new and long-time employees know how to react if fire occurs.

SPECIAL HAZARDS

As an inspector of the potential of fire in a plant or smaller area, you must consider two essential problems: the sources of ignition, and the materials that can burn. These are of major significance in the plastics manufacturing process.

Sources of heat include the operating temperatures, electrical equipment and wiring, static sparks, friction, matches, and smoking materials.

Molding and extrusion operations require temperatures of 300 to 650°F, depending on the plastic being processed. Because the upper temperatures are beyond the practical use of heat transfer fluids, electrical resistance heating is most frequently used, and sometimes controllers do not operate correctly and temperatures become excessive.

If plastic feedstock is allowed to remain within equipment, it may decompose under excessive or prolonged temperature and release combustible gases. This hazard may be reduced through frequent cleaning of molding and extruding areas.

The potential of friction sparks is always present when mechanical equipment is operating. If tools, sticks, or other objects accidentally fall into rollers, belts, or other powered moving devices, the consequence can be frictional sparking or other heating, perhaps severe enough to ignite combustible vapors.

Electrical wiring and connections are frequently subject to damage or misuse, and you, as inspector, should examine all wiring and installations for adequacy and safety.

Static sparks are common in plastic manufacturing because plastics are good electrical insulators. Sparks can be generated by the movement of film across rolls or guides, or by transmission belts. The hazard can be diminished by correct grounding of equipment and by the use of tinsel conductors on moving films or filaments.

The danger of smoking materials can be diminished by confining the use of such materials to certain nonhazardous rooms and areas and providing sufficient ash trays and other means of disposal. Warning signs should be displayed prominently in all areas where smoking materials would be a hazard.

Another potential source of ignition is in the hydraulic systems for clamping molds or providing pressure to rams or screws that force plastic by compression, transfer, or injection molding. Temperatures in this process may exceed the ignition temperature of some petroleum fluids.

FIRE BEHAVIOR

As an inspector, you should try to learn how plastics burn and emit products of combustion. However, it is not reassuring to learn that they can burn in many different ways and can produce mildly or severely toxic gases, or smoke ranging from nearly colorless to impenetrable black. Another disturbing fact is that reliable test methods on the relative hazards of materials have failed to predict the fire behavior of some plastics, but at least some features have been defined, as follows:

Ease of Ignition

Plastics may have a higher ignition temperature than wood, but some can be ignited with a small flame and burn quickly. Some plastics have a surface flame spread that is ten times that of wood, up to 2 feet per second.

Smoke

Some plastics emit great quantities of heavy smoke after they become ignited, and this dense, sooty smoke can overcome any person in the area who is not protected by adequate breathing apparatus. Even the smoke of lighter shade can be severely irritating and perhaps fatally toxic.

Toxic Gases

Any plastics fire will generate lethal quantities of carbon monoxide, and possibly hydrogen cyanide, hydrogen chloride, phosgene, and other products of combustion.

Dripping

A fire may cause thermoplastics to melt and drip into a small pool of liquid. The vapors from this liquid may ignite and flare to extend the fire, or the flames may flash up the tar-like dripping.

Corrosion:

Fires involving common plastics, such as polyvinyl chloride, have caused severe corrosion to sensitive electronic equipment and metal surfaces.

FIRE CONTROL

In most situations, water is the most appropriate extinguishing agent for fires involving plastics, but it should be available in large quantities. A plastics manufacturing plant should have automatic sprinklers, standpipe and hose systems, water-type portable extinguishers, and perhaps special automatic extinguishing systems for flammable liquids and electrical fires. Because of the many variables in plastics and their fire behavior, the arrangement of extinguishing and explosion suppression systems and portable equipment should be designed for each individual plant, even though fire experience has verified many practical fire protection measures. For example, there are flammable and combustible products and vapors that require suppression by inert gas or foam, rather than water.

The burning of plastics is influenced considerably by their

physical form. Molding pellets in bulk storage will burn differently from certain finished products, such as containers, polyvinyl envelopes, or insulated cables. Dust and some granules may flare rapidly on the surface of equipment, but a solid compound in a mold might be easy to extinguish.

If large quantities of plastics in storage are exposed to fire, then fire fighters should direct hose streams to cool these supplies and prevent them from melting into the more flammable liquid state. At the same time they must be careful not to agitate dusts of plastics or wood or other products.

When a fire involves plastics, the toxicity problem is so serious that special procedures are necessary to alert plant personnel and fire fighters of the hazard. It is not sufficient to rely on signs, signals, or a planned evacuation routine. All personnel who are not protected by breathing apparatus must leave the building and stay outside until the emergency is over. The fire officer in charge of responding fire fighters must be told immediately of the respiratory hazards and the need for protective breathing apparatus.

The amount of manual fire fighting will depend upon the severity of the fire and the capability of automatic fire extinguishing equipment. Fortunately, in most incidents, the fire is suppressed quickly with little damage or injury to personnel. This is the usual result in plants that have regular inspections and follow good practices of housekeeping and maintenance.

As inspector you must also become familiar with the components of automatic extinguishing systems so that you can verify their condition and functioning capability.

SI UNITS

The following conversion factor is given as a convenience in converting to an SI unit the English unit used in this chapter.

$$\% \, (°F - 32) = °C$$

BIBLIOGRAPHY

Fire Protection Guide on Hazardous Materials, 6th ed., National

Fire Protection Association, Quincy, MA, 1975. This publication contains the Flashpoint Index of Trade Name Liquids and four NFPA standards of great use to fire inspectors because they present detailed information on the fire characteristics and properties of combustible gases and solids and flammable liquids.

McKinnon, G. P. ed, *Industrial Fire Hazards Handbook*, 1st ed., National Fire Protection Association, Quincy, MA, 1979. A comprehensive description of industrial fire risks, life safety from fire, and general and specific fire hazards in a large variety of industrial manufacturing, processing, and assembly plants.

McKinnon, G.P. ed., *Fire Protection Handbook*, 15th ed., National Fire Protection Association, Quincy, MA, 1981. Section 8, Chapter 3 describes plastics fabrication plants.

NFPA Codes, Standards, and Recommended Practices. (See the latest NFPA *Codes and Standards Catalog* for the availability of current editions of the following documents.)

NFPA 69, *Standard on Explosion Prevention Systems*. A description of the equipment and systems needed to keep oxidant and combustible concentrations below hazardous levels; discusses the functions, inspection, and maintenance of explosion suppression systems and the requirements for inert gas systems.

NFPA 70, *National Electrical Code*. Contains requirements for all standard types of electrical installations.

NFPA 654, *Standard for the Prevention of Fire and Dust Explosions in the Chemical, Dye, Pharmaceutical, and Plastics Industries*. This document describes the classification and hazardous ratings of plastics, describes practical measures of building construction, area segregation and venting, methods of preventing explosions and of minimizing damage to equipment, and the fire protection needed for buildings and apparatus.

Chapter 21

EXPLOSIVES AND BLASTING AGENTS

If you are responsible for inspecting firesafety conditions in a plant that manufactures, stores, and transports explosives, you probably know the hazards of these materials and the importance of maintaining clean and safe conditions. If personnel are not continually alert to the need for safety, and if good housekeeping and fire protection principles are not practiced, a spark, a flame, an impact, or the decomposition of explosive materials can lead to catastrophic destruction. In addition, you and other personnel will be exposed to a greater potential of severe danger.

Your knowledge of the characteristics of these materials, your examination of the plant's areas, and your correction of hazardous situations are of major importance in the fire protection level of effectiveness. The general duties of an inspector are defined in Chapter 2 of this manual, but inspectors in plants or motor terminals where explosives are present must be concerned about combinations of chemicals and liquids that have a wide range of sensitivity, potential flame and explosive power. Your task is to find, identify, and correct or remove all factors that are dangerous. You must also know about the extent and efficiency of existing fire protection and the plans for controlling incidents of fire or explosion.

IMPORTANT TERMS

In order to understand the potential problems it is important to be familiar with the terms and definitions that apply to explosive materials. The following are brief samples digested from a much larger listing.

Blasting Agent. A material or mixture intended for initiating an explosion, but mixed or made to have sufficient insensitivity to electrostatic effects, impact, and exposure to heat.

Explosive. A chemical compound mixture or device whose primary or common function is to create an explosion.

Composite Propellants. A mixture of an oxidizer with elastomeric. Used in gas generators and rocket motors.

Detonating Cord. A flexible cord that contains a center core of high explosive. It is used to initiate other explosives.

Detonator. Any device for initiating detonation.

High Explosive Materials. Produce a high rate of reaction, high pressure development, and a detonation wave in the explosion.

Low Explosive Materials. Produce deflagration, or a low rate of reaction and pressure.

Oxidizing Materials. Any solid or liquid that yields oxygen or other gas that reacts to oxidize combustible materials.

Propellant. Functions by deflagration and normally is used for propulsion.

Primer. Package or cartridge of explosive material with a detonator, or detonating cord attached to a detonator. Used to initiate blasting agents or other explosives.

Sensitivity. Tendency of explosive material to detonate on receiving impact, heat or other influence which can cause explosive decomposition.

Special Industrial Explosive Materials. Shaped materials, sheet forms and other extrusions, pellets and packages of explosives used in metal fabrication and for reduction of scrap metal.

Water Gel. Any explosive or blasting agent that contains a substantial portion of water.

In addition to these and other technical terms, it is important for you to understand the classification of explosives as defined in the Hazardous Materials Regulations of the U.S. Department of Transportation. These are:

Class A Explosives: Possess detonating or otherwise maximum hazard, such as dynamite, desensitized nitroglycerin, lead azide, fulminate of mercury, black powder, blasting caps, and detonating primers.

Class B Explosives: Possess flammability hazards, such as propellants, including some smokeless propellants, and photographic flash powders.

Blasting Agents: Possess minimal accidental explosion hazard.

Class C Explosives: Include certain manufactured articles which contain Class A or Class B explosives, or both, as components, but in restricted quantities.

Forbidden Explosives: Explosives forbidden from or not acceptable for transportation by common carriers.

MIXING PLANTS AND VEHICLES

Certain principles of firesafety are common to every area of manufacture, transportation, and storage of explosive materials. Each area and its equipment must be clean and free of deposits of the materials; matches and other smoking materials cannot be permitted; buildings must be made of noncombustible materials, or of sheet metal on wood studs; floors must be of concrete or other nonabsorbent material; portable fire extinguishers and other appropriate equipment must be readily available and fully charged; and primers and detonators are kept separate from the explosive materials.

In addition to these principles, you, as inspector, will have to consider the less obvious weaknesses or violations of firesafety. First are the relative cleanliness and order of the building interior and the work stations. Check to see if aisle space and other passageways are clear of obstructions. Note whether there are accumulations of cloths, paper or other combustibles, and whether there are residues of flammable liquid or grease on equipment or the floor. Watch for accumulations of dust on walls and equipment. Check the heating equipment to verify that it does not produce flame or sparks inside the building.

Verify that portable fire extinguishers have been inspected and charged within the year and they are at their designated locations. Inspect the systems of automatic extinguishing or explosion suppression equipment to make certain that it is in good condition and ready to operate. Observe how unopened and emptied containers are stored and used and look into the emptied containers to see if they have residues of the hazardous products.

Examine all electrical outlets and operating machinery for possible misuse or overloading and check the main board for condition of fuses and circuit breakers.

Go to the storage and shipping areas and observe how containers are arranged and handled, and how vehicles and lifting equipment are operated. Watch particularly for containers or spills of oil or other flammable liquids.

These inspection procedures should be supplemented by your observations in the specific areas of manufacture, transportation and storage.

BLASTING AGENTS

Blasting agents are manufactured so that the final product is relatively insensitive, but the materials from which they are made have their own hazards, which require respect. Blasting agents consist of an oxidizer mixed with a fuel. Under fire conditions, oxidizers yield oxygen and vigorously support combustion. Ammonium nitrate, for example, is a compound produced by reacting nitric acid with ammonia, and the end product will be fertilizer grade, dynamite grade, nitrous oxide grade, or some other mixture. It is capable of detonating with about half the blast effect of explosives, if it is heated under confinement that permits pressure buildup, or is subject to strong shock, as from an explosion.

Here are some other precautions and items to consider in inspections of blasting agent mixing facilities:

Fuel oil storage must be outside the mixing plant and located so that oil will drain away from the plant if the tank ruptures.

The mixing building must be well ventilated. Check to make certain that emergency venting operates correctly.

Internal combustion engines used for generating electric power must be located outside the mixing building, or shielded by a firewall. Spark emission can be a hazard to materials in the plant.

Mixing and packaging materials must be compatible with the composition of the blasting agent.

Oxidizers are sensitive to heat, friction, impact, and impurities, and must be processed and stored accordingly.

The flash point of No. 2 fuel oil, 125°F, is the minimum permissible for hydrocarbon liquid fuel for the agent mix.

Metal powders, such as aluminum, are sensitive to moisture, and should be secured in covered containers.

Solid fuels, particularly of small size, create a dust accumulation problem. Dust can be removed by vacuuming with appropriate nonsparking equipment, or by washing.

There should be no drains or piping in the floor where molten materials can flow and be confined during a fire.

Empty ammonium nitrate bags must be disposed of daily in a safe manner.

The entire building must be cleaned thoroughly on a regular basis.

Do not permit smoking, matches, open flames, or spark-producing devices, or firearms within 50 feet of the building.

Make certain that the area around the plant is cleared of brush, dried grass, leaves, and other litter for at least 25 feet away from the building.

Explosives must not be stored within 50 feet of the building.

Explosives that are not in the process of being manufactured, transported, or used must be stored in appropriate magazines.

Ammonium Nitrate

Facilities for the mixing, handling and storage of ammonium nitrate need the same fire precautions as those using other oxidizers. Because this compound is sensitive to contamination and heat, you, as inspector, should be alert to these two influences.

Buildings in which this compound is stored must not exceed one story in height and should not have basements, unless the basement is open on one side. There should be adequate ventilation or automatic emergency venting.

All flooring in storage and handling areas must be noncombustible or protected from impregnation by ammonium nitrate. There must be no open drains, traps, tunnels, pits or pockets where the compound can accumulate in a fire.

Containers of ammonium nitrate must not be stored if the temperature of this agent exceeds 130°F.

Bags of this product must be stored at least 30 inches away from building walls and partitions. Storage piles should not exceed 20 feet in height, the same distance in width, and 50 feet in length, unless the building is of noncombustible construction or protected by automatic sprinklers. Storage piles must be at least 3 feet below the roof or beams overhead.

Aisles must be at least 3 feet wide with at least one service or main aisle at least 4 feet wide.

In bulk storage of ammonium nitrate, bins must be kept clean of any contaminating materials. Aluminum or wooden bins should be used, because this agent is corrosive and reactive in combination with iron, copper, lead, and zinc. The storage should be clearly identified by signs reading "AMMONIUM NITRATE" in letters at least 2 inches high.

It is important that this material not be stored in piles of excessive height. The pressure setting of the mass is affected by humidity and pellet quality as well as temperature. Temperature cycles through 90°F and high atmospheric humidity are not desirable for this product.

Ammonium nitrate can be affected by a wide range of contaminants, including flammable liquids, organic chemicals, acids, and other substances. It is important that these contaminants be kept out of the storage building or be kept at some distance and shielded from this agent according to technical requirements.

Electrical installations must conform with NFPA 70, *National Electrical Code*, for ordinary locations, and be designed to minimize corrosive damage.

Spilled materials and discarded containers must be removed and disposed of promptly.

Open flames and smoking must be prohibited from these storage buildings.

FIRE PROTECTION

Automatic fire protection, particularly sprinkler systems, should be provided in an area where ammonium nitrate is stored, handled, and processed. These should be supplemented by portable fire extinguishers, standpipe systems, and fire hydrants.

If fire starts in such a processing plant, it is important that the ammonium nitrate be covered by large quantities of water until the burning is extinguished. If the fire grows beyond control, all personnel should leave the area. If ventilation of the building is not accomplished automatically, it should be done manually if possible. The fire should be approached from upwind, because the vapors of burning ammonium nitrate are very toxic. Fire fighters should wear self-contained breathing apparatus, and other persons, not so protected, should be ordered to leave the area.

After the fire is extinguished, loose and contaminated ammonium nitrate should be dumped in water or buried, if that is acceptable to environmental authorities. The remainder can be dissolved, flushed, or scrubbed from all areas. Wet, empty bags and containers should be flushed, permitted to dry, then burned.

Remember these principles during inspections.

MOTOR VEHICLE TERMINALS

There are three common assembly points for trucks that

haul explosive materials: explosives interchange lots, explosives less-than-truckloads lots, and explosives motor vehicle terminals. At each of these places, large quantities of sensitive explosive materials may be brought into proximity and be subject to accidental fire or vandalism. Accordingly, each of these places requires certain fire prevention measures, much like the following:

Interchange lots should be separated at least 100 feet from other facilities. Weeds, underbrush, vegetation, and other materials should be cleared for at least 25 feet from the lot, and adequate warning signs should be posted. Fences, gates and security patrols are important backups to these measures.

Portable fire extinguishers should be placed in appropriate locations and hoses may be connected to hydrants and standpipes. Lots should be protected by natural or artificial barricades.

At least 5 feet should separate vehicles parked side by side or back to back; 25 feet between vehicles loaded with explosives.

Smoking, matches, open flames, spark producing devices, and firearms should not be permitted within 50 feet of the lot.

SI UNITS

The following conversion factors are given as a convenience in converting to SI units the English units used in this chapter.

$$1 \text{ in.} = 25.4 \text{ mm}$$
$$1 \text{ ft} = 0.305 \text{ m}$$
$$\frac{5}{9} \, (°F - 32) = °C$$

BIBLIOGRAPHY

McKinnon, G.P. ed., *Fire Protection Handbook*, 15th ed., National Fire Protection Association, Quincy, MA, 1981. Section 4, Chapter 7 contains data on storage distances and fire protection for explosives and blasting agents.

NFPA Codes, Standards, and Recommended Practices. (See

the latest *NFPA Codes and Standards Catalog* for the availability of current editions of the following documents.)

NFPA 70, *National Electrical Code*. This code includes the requirements for all electrical installations, including those that are essential for explosives mixing plants.

NFPA 490, *Code for the Storage of Ammonium Nitrate*. A code that covers the storage requirements for bags, drums, and other containers, bulk storage, contaminants, general safety measures, and fire protection needs.

NFPA 495, *Code for the Manufacture, Transportation, Storage and Use of Explosive Materials*. This code includes terminology, and chapters on security and safety, blastings agents, water gel and emulsions, transportation, aboveground storage, and other essentials.

NFPA 498, *Standard for Explosives Motor Vehicle Terminals*. This standard covers firesafety requirements for explosives interchange lots and less-than-truckload lots, with particular emphasis on vehicle parking, control of ignition sources, security against trespassers, and employee training.

Chapter 22

HEAT UTILIZATION EQUIPMENT

Heat is probably the major requisite for most industrial processes or products. It is used to provide energy for operating machinery or to change the form or character of raw materials to adapt them to some human desire or need. Heat is used in boiler furnaces to transform water into steam to drive the turbine generators which provide electric power. The electric power, in turn, is used for lighting, for operating machinery, or for initiating chemical reactions, such as that required in the smelting of aluminum. Heat is used in ovens, kilns, and dryers to cure, change, or preserve materials and products.

By its very nature, heat utilization equipment, in its many and varied forms and purposes, presents serious fire hazards. Inspectors must be familiar with the types and functions of heat utilization equipment so that they may recognize the associated hazards and see that the necessary maintenance and protective measures are taken to assure life safety and firesafety.

BOILER-FURNACES

Boiler-furnaces, or boilers, so they often are called, are used to turn water into steam. That steam, as mentioned, may be used to drive electric generators, one of its most important uses in industry. It may be directly used to power steam-driven machinery, or it may be used as process steam.

There are two basic types of boiler-furnaces: the water-tube and the fire-tube. In the first of these, water is circulated through tubes heated by hot combustion gases. In fire-tube boilers, the combustion gases pass through tubes immersed in circulating water.

Basically, the combustion of fuels in boiler furnaces involves carbon, hydrogen, and sulfur burned with oxygen from air. The heat released is about 14,000 Btu per pound of carbon and 61,000 Btu per pound of hydrogen. The sulfur plays little part in the production of heat, but it can be a major source of corrosion and pollution problems.

Oil, natural gas, and pulverized coal are the most common fuels for boiler furnaces. However, some industries may use waste products including gases, flammable liquids, wood waste and sawdust. These may require special precautions.

Figure 22-1. Installation for combination oil and gas firing in a large electric utility boiler.

Whatever fuel is used, the combustion process results from the continual introduction of fuel and air in a combustible mixture. Flow rates of fuel and air must be controlled as must the fuel-air ratio and the ignition source. If any of these are irregular or interrupted, a boiler explosion can occur.

Boiler operation requires different outputs to meet varying load conditions. The operating, or load, range of a boiler is the ratio of the full load to the minimum load at which it will operate reliably and produce complete combustion without changing the number of burners in operation.

For oil to be burned efficiently, it must be atomized. This is usually done by either steam or mechanical atomizers. Steam atomizers produce a steam-fuel emulsion that atomizes the oil by the rapid expansion of the steam when released into the furnace. Steam atomizers perform more efficiently over a wider load range than other types. Normally they atomize properly down to 20 percent of rated capacity.

Premix gas burners mix fuel and air before they are introduced into the burner-nozzle. External mix gas burners mix the fuel and air outside the nozzle. Multiple speed external mix burners provide high ignition stability and are replacing many other types.

A multifuel-fired furnace with multiple speed burners and proper control equipment can be changed from one fuel to another without a drop in load or fuel pressure. Simultaneous firing of natural gas and oil is acceptable with this type of burner.

There are several possible arrangements for pulverized coal-fired systems due to the many functions necessary. Coal must be transported to the pulverizer in measured and controlled quantities. The air-coal stream from the pulverizer must be maintained within a specific temperature range to increase efficiency and reduce the hazard of coking and fire. The coal-air mix is combined with a controlled amount of secondary air at the burner. The fuel must be completely consumed in continuous process with no more than a trace of combustibles in the stack gas and ash hoppers. The amount of oxygen in the initial combustion zone is limited to control stack emissions.

Dual register burners proportion the air between the fuel-rich ignition zone and the secondary combustion zone to ensure full combustion.

The principal hazards of boiler furnaces are explosion and fire. Explosions are the result of the ignition of combustible fuel-air mixtures that have accumulated in confined spaces. Loss of flame from an interruption in fuel or air delivery or in ignition energy may permit such an accumulation.

The accumulations generally are the result of equipment malfunction or operator error such as failure to purge the furnace between unsuccessful attempts to light off the burner.

OVENS AND FURNACES

Ovens and furnaces comprise a wide range of process heat equipment. Ovens are variously defined as chambers "used for baking, heating or drying," or "equipped to heat objects within." Furnaces are enclosures "in which energy in a non-thermal form is converted to heat by the combustion of a suitable fuel." A rule of thumb classifies ovens as heating devices which operate at temperatures below 1,400 °F. This does not always apply. Coke ovens have temperatures above 2,000 °F and some furnaces operate at a temperature below 1,400 °F.

The NFPA system classifies ovens and furnaces as follows:

Class A includes equipment operating at or near atmospheric pressure and in which there is a potential fire or explosion hazard due to flammable volatiles or combustible residue from objects processed. Flammable volatiles or residues are produced by paints, powders, or finishing processes such as dipping, spraying, coating, impregnation, polymerization, or molecular rearrangement.

Class B ovens and furnaces include equipment operating at approximately atmospheric pressure and for which there are no hazards from flammable or combustible volatiles or residues.

Class C furnaces are those in which there is an explosion hazard due to some flammable or other special atmosphere. This type may use any heating system and includes the atmosphere generator when one is used.

Class D furnaces, which are the subject of NFPA 86D, are vacuum furnaces operating at temperatures up to 5,000 °F and pressures below atmospheric. Such furnaces may use any type of heating system and may also use special atmospheres.

More complete information on the characteristics and uses of each of these types is found in Section 9, Chapter 5 of the *NFPA Fire Protection Handbook*, 15th Ed.

Ovens and furnaces also are designated as batch or continuous type. In the batch type, the temperature is constant. The material is introduced and remains in place until the process is complete and is then removed, usually by the opening through which it entered. In the continous type, the material moves through the furnace on some type of conveyor. The oven temperature may be constant, or it may be divided into varying zones.

The most common methods of transferring heat to the materials being processed are:

- Direct contact with the products of combustion.
- Convection and radiation from the hot gases.
- Reradiation from the hot walls of the furnace.

Oven and furnace heaters are of two types: direct-fired and indirect-fired. Direct-fired ovens heat by contact with the products of combustion. Indirect-fired heat by radiation from heated tubes of air. Indirect-fired ovens are safer because dangerous fuel-air mixtures do not readily fill the enclosure. However, explosions are possible due to vapors from the drying process.

There are several different arrangements for the firing, which may be either internal or external, direct or indirect.

Many different sources of heat may be used for ovens for furnaces. These include gas and oil burners, electric heaters, infrared lamps, electric induction heaters, and steam radiation.

Ovens and furnaces should be located where they will present the least possible hazard to life and property. They may need to be surrounded by walls or partitions. The area should be adequately ventilated and have proper explosion venting. Operating and control equipment should be tested regularly and all portions of the oven or furnace and its attachments should be cleaned on a regular schedule.

Atmosphere generators provide the special gases required for some heat utilization processes. Generally these atmospheres are toxic, flammable, or explosible. Exothermic generators produce the gas by partially or completely burning fuel gas at a controlled ratio, usually 60 to 100 percent aeration. Endothermic generators use a ratio of less than 50 percent. Ammonia dissociators, by temperature reaction with a catalyst, produce dissociated ammonia (25 percent hydrogen and 75 percent ammonia) from ammonia. Special atmosphere generators must have adequate supervisory controls. NFPA 86C provides information on protective equipment for atmosphere generators.

AFTER-BURNERS AND CATALYTIC COMBUSTION SYSTEMS

After-burners and catalytic combustion systems are used to conserve oven fuel and reduce fumes, odors, vapors and gases to acceptable exhaust products such as carbon dioxide (CO_2) and water vapor. Some exhaust from ovens and furnaces may require special treatment to remove particulates, halogens, hydroxides, and sulfur and nitrogen oxides.

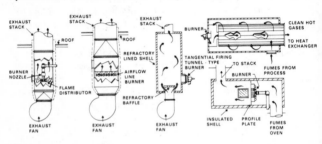

Figure 22-2. Typical direct flame fume incinerators. (Maxon Premix Burner Co., Inc.)

After-burners (direct-flame incinerators) burn organic solvent vapors, organic dusts, and combustible gases. The fumes must be heated to autoignition temperatures and there must be sufficient oxygen to complete the chemical reaction. More than 16 percent oxygen is required to keep the fumes at or less than 25 percent of the lower explosive limit (LEL) or 50 percent if there are adequate combustible gas analyzers and interlocks.

Operating temperatures in the combustion chamber are usually between 1,200 and 1,500°F. Conversion to carbon dioxide (CO_2) at 92 percent efficiency at 1,300°F, and at 96 percent at 1,450°F, has been reported in tests.

After-burner combustion chambers may be either lined with heavy refractory materials and have external burners or lined with light refractories and equipped with line burners.

If the fumes to be incinerated are inert gases with low combustible content, they may be mixed with sufficient air and burned. If they are combustible and concentrated from 25 to 100 percent of the LEL, they should be diluted with air before transfer to the incinerator. Combustible fumes above the LEL are normally burned in flare stacks or as fuel in heating equipment, a use which requires special equipment and control.

A catalytic combustion system uses a catalyst to speed up combination of fuel-air or fuel-gas mixtures. Catalytic heaters may be used to burn a fuel gas with much of the energy released as radiation to the processing zone. Or they may be used in the oven exhaust to release heat from evaporated by-products to a heat exchanger.

Catalytic combustion elements are of three types.

• An all-metal mat for use as a fuel-fired radiant heater or to oxidize combustibles in fume-air mixtures.

• Ceramic or porcelain elements using rare earths, platinum, or metallic salts as catalysts.

• A bed with pellets or granules retained between screens but with individual pellets or granules free to migrate within the bed.

HEAT RECOVERY

Heat exchangers and direct recirculation of heat are often employed to make process and fume incineration more economical. Such heat recovery, when applied to heat generation and fume incineration, can supply a significant portion of the heat requirements. The recovered heat may be used for:

• Primary or supplementary source of process heat.
• Some process zones in a multi-zone process.
• Other nearby processes.
• Pretreating fumes to incinerators.

- Heating make-up air.
- A waste heat boiler servicing multiple requirements.

Combustible deposits and flammable liquids heated to high temperatures within the heat exchanger present fire and explosion hazards. Dirty stream deposits may make it inoperable.

LUMBER KILNS

Lumber kilns find their value in the fact that they can dry freshly cut wood to a usable moisture content more quickly and more efficiently than can be accomplished by outdoor or natural seasoning. Natural drying often results in wastage and loss due to excessive warping and cracking.

Though often called dry kilns, wood dryers usually employ moisture to maintain a uniform, but lessening, content during the drying to reduce warping, checking, and cracking.

Lumber kilns may be classified in three ways:

- By the method of heating.
- By the method of air circulation.
- By the method of operation.

In batch or compartment kilns, the lumber remains stationary during the entire process. In continuous kilns, lumber enters at one end and, over a period of time, progressively moves to the discharge end.

Kilns may be directly or indirectly heated. In the first type, hot gases produced by burning oil, gas, sawdust, or other fuel are passed through the stacked lumber. Direct heating may also be done by using open oil or gas flame to heat large metal surfaces, which act as heat exchangers.

Steam, circulated through pipes located at either the bottom or top of the kiln, is a common source of heat for indirectly heated kilns. Hot gases circulated through ducts and electrical resistance heaters also are used.

In natural circulation kilns, heated air rises up through the stacked lumber. As it cools it travels down to the heater and is reheated. During the first part of the drying cycle, the air moves upward at the sides of the kiln and down through passages within the stack. When the moisture has been reduced to around 10 to 20 percent, heated air is directed upward through the center. Vents in the wall or roof exhaust the air and moisture.

In forced circulation kilns, air is moved over and through the stack by blowers, either internal or external. Heating elements and blowers may be located either at the bottom or top of the kiln.

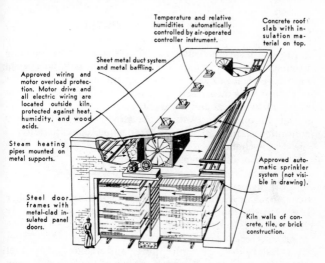

Figure 22-3. A forced-circulation double track compartment kiln. Note that automatic sprinklers are installed above and below the platform between the kiln and the overhead fan room.

Though the air-flow systems in batch and progressive kilns are much the same, there is one minor difference. In progressive kilns, the air moves from the dry end to the green end. Because the air has lost heat and picked up moisture, the rate of drying is slower at the green end.

Lumber kilns present serious fire hazards, especially when direct-fired or high pressure steam systems are used and when the structure itself is combustible. Direct-fired kilns are analagous to Class A ovens and require the combustion controls specified for drying ovens where the heating fuel is introduced into the oven itself.

Kilns should be of fire resistive or heavy timber construction. They are subject to extreme variations in temperature and humidity and to unusual expansion and contraction which reduce structural stability.

DEHYDRATORS AND DRYERS

In some respects, dehydrators and dryers for agricultural products and lumber kilns are similar. The basic purpose of each is to remove moisture from an organic material. Kilns introduce some moisture into the process only for control purposes. Dehydrators and dryers do not.

Dehydrators and dryers are classified as batch, bulk, and continuous. They differ in the arrangement and operation of the drying chamber.

Continuous dryers include:
- Drum dryers for milk, puree and sludge.
- Spray dryers for milk, eggs and soap.
- Flash dryers for chopped forage crops.
- Gravity dryers for small grains, beans and seeds.
- Tunnel dryers for fruits, vegetables, grains, seeds, nuts, fibers and forage crops.
- Rotary dryers for milk, puree and sludge.

Tunnel dryers may also be batch dryers and may be further classified according to airflow and whether or not intermediate heating is required.

Batch dryers may be either fixed or portable. They may be gravity type. They also include pan dryers for sugar, puree, sludges and other products.

Bulk dryers are used to dry seeds, grains, nuts, tobacco, hay, and forage in the bin, crib, or compartment in which the product is to be stored. Such dryers commonly introduce dry, heated air below a perforated floor. The air rises through the product, carrying moisture to exhaust vents.

In batch dryers, heated air is introduced into a perforated plenum. The product to be dried is fed from overhead and loses moisture as it passes downward. Exhaust air escapes through openings in the outer wall of the dryer.

In continuous gravity dryers, wet material enters at the top of a silo-like structure. Warm air is blown into the upper half to dry the falling material, which continues through a zone of cool air to the bottom where it is conveyed to storage or a process step.

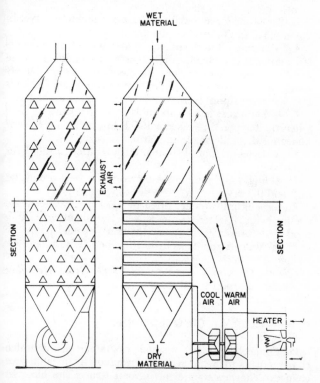

Figure 22-4. A continuous flow-type gravity dryer.

Agricultural products dryers may be direct-fired or indirect-fired. They may be heated electrically or by a heat transfer medium such as steam. Heaters may be oil-fired, gas-fired, or solid-fuel-fired. When gas-fired, infrared heaters, or lamps are used, the focal length should be such that the surface of the product does not reach ignition temperatures. Electrical infrared lamps should be located where they cannot collect combustible dust.

Excluding those required for burners and heating equipment, dryer controls should include:

• A method for automatically shutting down the dryer in the event of fire or excessive temperature.

• A thermostat in the exhaust system when the product is fed automatically from the dryer to a storage building. In the event of excessive temperature, the thermostat should shut off heat to the dryer; stop the airflow except when the product is in suspension; stop the product flow; and sound an alarm.

• A thermostat in a combustible dryer that shuts off heat when the temperature reaches 165°F but permits a flow of unheated air, and sounds an alarm.

• A device to shut off heat if airflow through the dryer stops.

• A high-limit thermostat between the heat-producing unit and the dryer.

An agricultural product dried by heat requires adequate cooling before it is stored or packaged. The amount of cooling needed to prevent subsequent ignition depends upon the material and its spontaneous heating and how it is to be packaged or stored.

NFPA 61B, *Grain Elevators and Bulk Grain Handling Facilities*, has specific recommendations for grain dryers.

OIL QUENCHING

Oil quenching is a method of imparting desirable characteristics to metals and metal products by metallurgical changes. The process presents several serious fire hazards, chief of which are the combustible character of the quenching oils and the elevated temperatures often required.

Other contributors to the hazards are:

- Special atmospheres.
- Size and shape of the work being processed.
- Oil volume.
- Location of furnaces and quench tanks.
- Exposure to other process or storage facilities.

The most critical step in the process is the entrance of the work into the quench, which must be rapid and complete. There must be minimal splashing and no overflow of the quench oil. Partial immersion is the cause of most quench oil fires.

Chutes, elevators, cranes, hoists, conveyors, or a combination of these, move the work into, through, and out of the quench. They must be designed and maintained properly, so that parts or products move without jamming. Baskets and elevators cause more partial immersions than does any other method. Elevating mechanisms must be adequately supported to prevent tilting of the load. Proper guides within the tank are necessary to ensure uniform movement and to prevent wedging, which could result in partial immersion.

Provision must be made to drain the work at the end of the quench cycle or an excessive amount of oil will be wasted. Usually the work is still warm at this time and oil may vaporize and be susceptible to ignition. Vapors will condense on surfaces and add to the seriousness of any fire at this point.

Quench oil must be maintained within certain specified temperatures. Generally this requires a cooling system to prevent overheating. A cooling system failure that permits water to enter the quenching oil is particularly hazardous. Hot work entering the bath can convert the water to steam and cause a boilover. Quench oil should be tested regularly. If the water content reaches 0.35 percent by volume, the oil is no longer safe to use.

The quantity of oil in the quench tank must be carefully controlled. If the oil level is too low and a large load is immersed, the oil can overheat and ignite. Too much oil will result in an overflow. The distance between openings in the tank wall and the liquid level with a full load submerged should be at least 6 inches. Adequate, fully trapped drains should carry overflow to a safe location or into special tanks.

Bottom drains should be provided for emptying a quench tank under serious fire conditions. This may be done either by gravity drains or special pumps so sized that the oil can be drained within 5 minutes. Such drains must be used only by well-trained persons, as improper use can result in greater hazard. If a flammable gas atmosphere is maintained above the oil, removing the oil can create a negative pressure that can result in explosion or greater fire severity. Established drain pipe sizes are listed in the NFPA *Fire Protection Handbook*, 15th Ed., Section 9, Chapter 8.

All automatic shutdowns should result in the workload being completely immersed or completely removed from the quench. All safety controls and interlocks should be tested on a regular schedule.

INSPECTION REMINDERS

- Look for evidence of fuel or hydraulic fluid leaks.
- Is there sufficient insulation or clearance where ducts or stacks pass through combustible walls, floors, or roofs?
- Are interlock safety systems intact?
- Look for evidence of corrosion, especially where high sulfur content fuel is used.
- Is the natural or mechanical ventilation adequate to remove flammable vapors?
- Have provisions been made for explosion venting where needed?
- Good maintenance procedures for heat utilization equipment include regular and frequent checks of safety devices and circuits. Ask to see the inspection reports and note any abnormalities that may signal a potential hazard.

SI UNITS

The following conversion factors are given as a convenience in converting to SI units the English units used in this chapter.

$$1 \text{ Btu/lb} = 2.326 \text{ kJ/kg}$$
$$1 \text{ in.} = 25.4 \text{ mm}$$
$$\frac{5}{9} (°F - 32) = °C$$

BIBLIOGRAPHY

McKinnon, G. P. ed, *Fire Protection Handbook*, 15th ed., National Fire Protection Association, Quincy, MA, 1981. Chapters 1, 3, 5, 6, 7, and 8 of Section 9 deal in detail with the equipment covered in this chapter.

McKinnon, G. P. ed., *Industrial Fire Hazards Handbook*, 1st ed., National Fire Protection Association, 1979. Several chapters in Part Three of this handbook discuss the equipment and hazards that are the subject of this chapter.

NFPA Codes, Standards, and Recommended Practices. (See the latest *NFPA Codes and Standards Catalog* for the availability of current editions of the following documents.)

NFPA 30, *Flammable and Combustible Liquids Code*. Covers piping, valves, and fittings used for flammable liquids.

NFPA 31, *Standard for the Installation of Oil Burning Equipment*. A standard on stationary and portable equipment, tanks, piping, and accessories.

NFPA 34, *Standard for Dip Tanks Containing Flammable or Combustible Liquids*. Discusses coating, finishing, and treating processes.

NFPA 54, *National Fuel Gas Code*. Criteria for the installation, operation, and maintenance of gas piping and appliances.

NFPA 58, *Liquefied Petroleum Gases*. General provisions for LP-Gas equipment and appliances.

NFPA 68, *Explosion Venting Guide*. Fundamentals of explosion venting.

NFPA 70, *National Electrical Code*. Includes electrical requirements in hazardous atmospheres.

NFPA 85, *Standard for Prevention of Furnace Explosions in Fuel Oil- and Natural Gas-Fired Single Burner Boiler-Furnaces*. A standard for single burner watertube and firetube boiler-furnaces rated at 10,000 pounds of steam per hour and above.

NFPA 85B, *Standard for the Prevention of Furnace Explosions in Natural Gas-Fired Multiple Burner Boiler-Furnaces*. Covers the design, installation, and operation of these boiler-furnaces.

NFPA 85D, *Standard for the Prevention of Furnace Explosions in Fuel Oil-Fired Multiple Burner Boiler-Furnaces*. Outlines equipment, interlock, and alarm requirements of these systems.

NFPA 85E, *Standard for Prevention of Furnace Explosions in Pulverized Coal-Fired Multiple Burner Boiler-Furnaces*. A standard for design, installation, and operation of coal-burning systems.

NFPA 86A, *Standard for Ovens and Furnaces: Design, Location and Equipment*. A standard for Class A ovens or furnaces.

NFPA 86B, *Standard for Industrial Furnaces: Design, Location and Equipment*. A standard for Class B furnaces.

NFPA 86C, *Standard for Industrial Furnaces Using a Special Processing Atmosphere*. A standard for Class C furnaces.

NFPA 86D, *Standard for Industrial Furnaces Using Vacuum as an Atmosphere*. A standard for Class D industrial furnaces.

Chapter 23

FINISHING OPERATIONS

If you inspect a spray booth, room, or area, or a building designed specifically for finishing operations, you can expect to find a number of common hazards, and some that are peculiar to this specialized work. Flammable liquids, combustible materials, electrical equipment, sources of sparks, accumulations of potentially explosive dust or powder, machinery, and an atmosphere that is contaminated by particulates are some of the common fire hazards.

High speed discharge of flammable and combustible products, overspraying, ignitible vapors, open floor spraying, dip tanks, powder application, and bulk storage of hazardous products are some of the special hazards you will find in finishing plants.

Because these plants are susceptible to fire and explosion, you probably will find portable and fixed automatic fire protection equipment and a personnel force that is well aware of the dangers.

Therefore, when you are about to begin an inspection, you will need to be informed about flammable and combustible liquids, spray application of finishing products, dip tanks, explosion venting, explosion prevention systems, static electricity, electrical installations, fire extinguishing systems, the fire hazards of materials, and at least a few other subjects. That is a challenging list of information to absorb, but your task is made easier by the experience of fires and explosions in these occupancies that has already been recorded, and the numerous technical standards, guides, and other references that are available.

TYPES OF COATINGS

Three kinds of finishing operations are described in this chapter: spray finishing, powder coating, and dipping. They are in the category of "special process fire hazards" because each involves the use of flammable or potentially explosive materials in hazardous conditions. The objects being coated may be used for a specific purpose — protective, decorative, lubrication, adhesive, structural, or for other purposes — but the hazards are in the finishing process.

Spray Finishing

This is accomplished by discharging atomized finishing products through a hand-held spray gun, a similar gun held by a machine, an airless atomizer, and by use of an electrostatic disc or bell.

The air spray guns feature a valve to control the flow of fluid, a valve to control the flow of air, and an atomizing nozzle. The nozzle is designed to allow the fluid and compressed air to impinge and create a spray of fine droplets. The spray pattern may be controlled by a separate air control valve.

Airless atomizers are also used as hand-held or machine-mounted guns, but they do not use compressed air. Instead, the coating liquid flows at high pressure through a nozzle and discharges a high velocity, paper-thin film. The nozzle pressures range from 300 to 3,000 psi, and discharge ranges from a few ounces to a few gallons per minute. Such nozzles are made of extremely hard material, such as tungsten carbide, to resist the wear of such flows.

There are also air or airless spray guns that are designed to apply the finishing product electrostatically. For this, the gun receives high voltage input, from 35,000 to over 100,000 volts with very low current — below 100 microamperes. The charged, atomized particles are attracted to the grounded conductive or conductive-coated workpiece and adhere to the surface.

A "hot spray" can be used in each of these processes. For this, the coating fluid is heated to some temperature between 100 and 200 °F, which reduces its viscosity and required air or fluid pressure, thus reducing the amount of overspray.

Another atomizer is the electrostatic disc or bell. This is

usually a disc of 6- to 12-inch diameter, mounted with its axis vertical, and charged electrically to about 100 kilovolts. As the disc spins, coating fluid flows onto its surface, spreads in a thin film under centrifugal force, moves to the sharp edge of the disc where it is disrupted by electrostatic forces, and then sprays in a 360 – degree circumferential pattern. Workpieces pass by on a conveyor belt and receive the coating. The disc is rotated by an electric motor or air-driven turbine.

Figure 23-1. Electrostatic Disc. (Ransburg)

Similar results are obtained by using a cup or bell with a diameter of 2 to 4 inches, which may be used in a hand-held gun or be machine-mounted.

The hazards that should be obvious in such operations include the discharge of a fine spray of flammable liquid; the possible accumulation of overspray; the possibility of static sparks or electrical faults; and the flammable atmosphere.

Your inspection should include defining the properties of the spray fluid; learning how often the overspray is removed; verifying the effectiveness of the bonding and grounding of the gun and conveyor; examining the electrical installations; and determining the capabilities of the power venting system.

Powder Coatings

Another means of coating products is by application of dry powder. This may be accomplished electrostatically or by heat. In the electrostatic process, the powder is suspended in air, charged from a direct current power source, then directed onto the workpiece where the coating of fine powder is held by electrostatic force. Then the coated workpiece is conveyed through a process oven which melts the coating film.

Hand-held or machine-mounted spray guns can be used for electrostatic powder coating, which has the benefit of

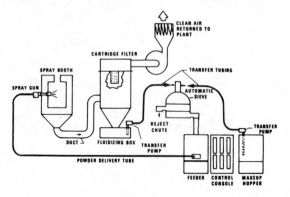

Figure 23-2. Diagram of automatic recycle of a single color powder coating system showing the booth, collector, fan, filter. (Nordson)

overspray recovery. Dust collection devices are used for this purpose, whereas in the fluid coating process the overspray must be discarded.

Another form of application is with an electrostatic fluidized bed in which the powder is held in suspension by air pressure. The upper layer of powder is charged electrostatically. When electrically grounded workpieces pass above the bed surface, they collect a deposit of powder. Then they pass to a curing oven.

These fluidized beds range in size from a small container up to larger than 8 feet across and 15 feet deep. When the airstream is adjusted properly, the powder inside the bed acts as a liquid.

Another method features a cloud chamber through which workpieces pass on a conveyor. Inside the chamber, apparatus blows powder into a billowing cloud, where the particles become charged electrostatically, then are attracted to the workpiece. After that the piece goes to the curing oven.

The powders used in this coating process are considered as ordinary combustibles, but there are some organic powders which can burn or explode when in suspension. Your inspection should include identifying the characteristics of the powder, the bonding for preventing static accumulation, the accumulation of powder overspray, the amount and use of powder containers, and storage practices.

Dipping and Coating

In these processes the workpieces are dipped in, passed through, or coated by flammable or combustible liquids. Equipment used includes dip tanks, roll coaters, curtain coaters, and flow coaters.

Dip tanks are containers of any size or shape, but they must be designed according to the degree of fire hazard. They should be constructed of noncombustible material to retain the liquid used in the process, and with consideration of corrosion, mechanical damage, maintenance and repair.

The dipping process consists of conveying the workpiece to

the tank, immersing it in the fluid, then removing it and passing it over a drainboard where excess liquid will flow back to the tank. The obvious hazard is the accumulation of flammable or combustible vapors in the atmosphere.

Flow coaters feature nozzles or slots through which the coating fluid is applied onto the workpiece as it is conveyed through a drip tunnel. The tunnel is enclosed on all sides, except for the conveyor opening. The excess liquid drops to a trough or sump.

Curtain coaters pump the coating material from a container or reservoir to a small reservoir above. A dam or weir is on one side of this head. When the coating material overflows this dam, it becomes a continual stream that drops to coat the workpiece. Excess coating drips into a trough and returns to be recirculated.

Roll coaters apply coating by direct contact to the workpiece by one or more liquid-coated rollers.

AREAS OF OPERATION

Spray coating application may be accomplished in a spray booth, a spray room or a spraying area, and fire protection arrangements are needed for each of these.

Spray booths may be the dry type or the waterwash spray type. The dry type does not have a washing system but does have distribution or baffle plates to facilitate air movement, overspray filters and rolls, and powder collection systems.

The waterwash type has a system for minimizing dusts or residues in the exhaust ducts and recovering overspray material.

A spray room is a fully enclosed, power ventilated room used exclusively for open spraying of flammable or combustible materials. The entire room is a spray area.

A spray area is a place where dangerous quantities of flammable vapors or mists, or combustible residues, dusts, or deposits, are present due to spraying operations.

HAZARDS

Fluid Spray Hazards

The hazards associated with fluid spray operations include the characteristics of flammable and combustible coatings, the vapors, and overspray; ignition sources; container use and storage; the transfer of flammable liquids; overused and contaminated filters; static sparks or buildup; misused or deteriorated electrical circuits and equipment; poorly trained and instructed personnel; inadequate ventilation and fire protection equipment.

Your inspections, therefore, will be easier if you become informed on each of these subjects. Sources of information on the fire characteristics of materials are listed at the end of this chapter.

You will have to look carefully for ignition sources because they may be too obvious to be considered. Smoking materials should not be permitted in the entire spraying area; a separate room or other location should be designated for that purpose. Similarly, cutting and welding or soldering operations should not be permitted until the entire area has been cleaned and vented and the cutting equipment can be isolated with adequate shielding or partitions.

All electrical wiring and equipment must be examined and compared with code requirements. Fixtures above hazardous areas must be totally enclosed or provided with a guard to prevent hot particles from dropping into the hazardous space.

Static sparks and electrostatic ignition can be prevented by bonding together all elements of the fluid transfer or spray system, and grounding the bond. Persons using the spray gun manually should also be bonded and grounded. There should be electrical interlocks in the system to de-energize electrical power supply when spray guns are not in use.

Vapors and overspray can be minimized by the use of barriers or enclosures and correct power ventilation. As inspector, you should verify the ventilation system capacity and flow rate performance.

Small quantities of flammable and combustible liquids may be brought to a mixing room where they can be prepared for use. Any flammable liquid containers larger than 5-gallon capacity that are kept indoors must have a plug with a pressure relief valve, a vacuum relief valve, and a flame arrestor. Bulk supplies of flammable liquid should be stored outside, away from buildings.

Prepared coating materials are transported from the mix room to spray areas through piping, or in containers that have tightly clamped lids.

During your inspections, always consider the housekeeping as an indication of attention to firesafety. There should be a consistent program of removing overspray residue and other materials from the interior of the spray booth, the conveyors, and the exhaust ducts. Blocked or contaminated filters should be replaced, then removed from the building because they may react chemically, or ignite spontaneously.

Personnel who work in the spraying area should know what to do if equipment malfunctions, or fire occurs; how to use portable fire extinguishers; what hazards exist in the spray materials; and what can be expected when automatic fire protection equipment operates.

Powder Coating Hazards

Because of the dangerous potential of flammable powders and dusts, the spray room or area should be made safe from all sources of ignition, similar to those listed for fluid spray operations in the preceding pages of this chapter. Mechanical sparking caused by one workpiece ingested into the ductwork leading from a spray booth was indicated as the cause of one collector explosion.

Powder collectors, flame detectors, and fast-acting damper systems are used to control and extinguish fire between a spray booth and powder collector. Automatic sprinklers and inert gas systems are also effective for spray operations. All such equipment should be examined in your inspections.

Dip Tanks and Coating

Because of the large quantities of flammable liquids used in these processes, the potential for ignition is high. Flammable vapors are not only present in the vicinity of the tanks, but are heavier than air and may flow along floors and into low recesses where they can be ignited by a spark or open flame. For this reason, one of the protective actions is to diminish the concentration of vapors to below the lower flammable limit (LFL). This is accomplished with properly designed ventilation and exhaust systems. In addition to removing common sources of ignition it is necessary to use equipment designed to be nonsparking, particularly electrical equipment. In fact, electrical installations probably deserve your first attention when you inspect dipping facilities, because extension cords and improvised connections are common hazards. Open flames and heated surfaces, static sparks, and heated equipment are other problem sources.

Exhaust systems of ovens and dryers should be interlocked with ventilation systems to provide immediate shutdown if the ventilating fails.

Static charges can also accumulate on the conveyors, rollers, collectors, festoon dryers, and other devices. These charges must be minimized by proper bonding and grounding.

Conveyor systems should stop automatically if the exhaust venting fails, or if a fire occurs.

Excess temperature controls are needed to keep flammable liquids below autoignition temperature. Controls are also needed to limit the surface temperature of heated workpieces.

The vicinity of dipping and coating operations must be kept clear of combustible residues and unnecessary materials. If excess residues develop, the operations should be stopped until the situation is corrected.

Depending upon the size and nature of the process, inspections should be made at least monthly to examine the equipment, covers, overflow pipe inlets and outlets, discharge, bottom drains and valves, electrical wiring and equipment, grounding and bonding connections, ventilation equipment (manual and automatic), and extinguishing equipment.

Dip tanks usually have a self-closing cover that is held open by a cable and fusible link. If fire occurs and the link melts, the cover drops and the fire is smothered.

Automatic sprinklers and foam, carbon dioxide, dry chemical, and halogenated agent systems may be used for protection, but portable fire extinguishers and other manually operated equipment are also necessary.

Personnel who work in dipping and coating operations should be trained for fire emergencies and understand the procedures required.

TIPS ON SAFE SPRAY OPERATIONS

Among general precautionary measures, additional to points mentioned elsewhere in this chapter, are the following:

Allow only nonsparking cleaning tools in the spray room area.

Have air-driven tools available for maintenance work.

Permit no welding or cutting in the area while the spray room is in operation.

Make daily checks of electrical motors in spray room equipment to see that they are not overheating and that fan blades are in alignment. Daily check the overcurrent protection of electric wiring, guards and globes on electric fixtures.

Inspect pressure hose and couplings periodically.

Make certain that the ventilation system is operating efficiently.

SI UNITS

The following conversion factors are given as a convenience in converting to SI units the English units used in this chapter.

$$1 \text{ in.} = 25.4 \text{ mm}$$
$$1 \text{ ft} = 0.305 \text{ m}$$
$$1 \text{ psi} = 6.895 \text{ kPa}$$
$$\tfrac{5}{9} \, (°F - 32) = °C$$
$$1 \text{ gal} = 3.785 \text{ litres}$$

BIBLIOGRAPHY

McKinnon, G.P. ed., *Fire Protection Handbook*, 15th ed., National Fire Protection Association, Quincy, MA, 1981. Section 9, Chapters 9 and 10 deal with spray finishing, powder coating and dipping.

McKinnon, G. P. ed., *Industrial Fire Hazards Handbook*, 1st ed., National Fire Protection Association, Quincy, MA, 1979. Contains considerable detail on spray finishing, dipping, and coating operations and equipment.

NFPA Codes, Standards, and Recommended Practices. (See the latest *NFPA Codes and Standards Catalog* for the availability of current editions of the following documents.)

NFPA 10, *Standard for Portable Fire Extinguishers*. Explains the sizes and capabilities of portable fire extinguishers, their testing, and their suitability for use on various kinds of fires.

NFPA 13, *Standard for the Installation of Sprinkler Systems*. Describes how such systems should be installed for maximum effectiveness, and discusses the requirements for hazardous locations.

NFPA 30, *Flammable and Combustible Liquids Code*. Describes the characteristics and fire properties of these liquids, requirements for processing buildings and liquid handling, fire control, sources of ignition, and handling.

NFPA 33, *Standard for Spray Application Using Flammable and Combustible Materials*. Outlines practical requirements for spray application with fluids and powders, for spray booths and rooms, ventilation, storage and handling, electrostatic apparatus, and related equipment.

NFPA 68, *Guide for Explosion Venting*. Explains the need for explosion venting of hazardous areas and discusses the required automatic equipment.

NFPA 69, *Standard on Explosion Prevention Systems*. Describes the uses of these systems and their functions in suppressing explosions of hazardous processes and storage.

NFPA 70, *National Electrical Code*. Defines the requirements for electrical installations and the kinds of hazardous areas that require nonsparking equipment and other safety measures.

NFPA 77, *Recommended Practice on Static Electricity*. Explains the hazards of static accumulation and the means of diminishing or eliminating those hazards. Bonding, grounding, and other measures are defined.

Chapter 24

WELDING AND CUTTING

It has been reported that welding and thermal cutting operations have caused as much as six percent of U.S. industrial fires in an average year of fire experience. This should not be a surprise to experienced fire inspectors who have seen the hazard potential and the dangers of these common operations.

Consider the normal activities and supplies required for these tasks: the use of oxygen and fuel gas burning at temperatures of 4,000 to more than 5,000 °F; an outburst of thousands of molten metal sparks; brilliant flare of light that can be harmful to unprotected eyes; high potential for physical burns or injuries; containers of fuel gas and oxygen; the usual amounts of tools and equipment that would be a hindrance to movement in an emergency; a small or large amount of combustibles that are not necessary in the area. These items together can create extreme fire danger, but separately they can be removed or eliminated in the interest of firesafety.

As an inspector you should understand how welding and cutting operations should be performed with efficiency and safety and how the major hazards can be minimized or eliminated.

PROCESSING METHODS

There are two basic processes of welding and thermal cutting: by electrical arc and by the heat of a fuel gas combined with oxygen. The electrical processes include shielded metal, gas metal, fluxed core, gas tungsten, and plasma arc welding, resistance welding, flash welding, and electroslag welding. Electrical cutting is achieved by air carbon arc or plasma arc process.

Oxy-fuel gas welding includes brazing and braze welding,

267

hard soldering, surfacing, certain heating operations, and other applications. Oxy-gas cutting can be done with portable equipment or a machine.

Arc Welding

In this work an electric arc is used to melt and join two metals. The arc is created between the metals to be welded and an electrode which is moved along the joint or held stationary while the workpiece passes beneath. Usually a filler metal is used to secure the joint, but not always. The electrode may or may not be consumed in the process, depending upon the kind used. Consumable electrodes provide the filler metal by melting, and its covering or core may shield the weld zone from atmospheric effects.

Shielded metal arc welding is used with ferrous-based metals and makes them coalesce by heating them with an arc between a covered metal electrode and the workpiece. This technique can be used in remote, unusual, or confined locations and is popular in construction, shipbuilding, pipeline work, and maintenance. It requires alternating or direct current power supply, power cables, and an electrode holder.

Gas metal arc welding uses a solid wire filler and a gas to shield the arc and metal. The filler serves as one terminal of the arc. The gas may be argon, carbon dioxide, or a mixture of these with helium or oxygen.

Fluxed core arc welding uses electrodes with a core of minerals and alloys, rather than a solid type. Some of these are intended for use with carbon dioxide-rich gases; others produce their own gaseous shielding.

In gas tungsten arc welding, argon and helium may be used as shielding gas, the electrode is nonconsumable tungsten, and a filler metal may be used. Equipment includes a power supply, welding torch, inert gas supply, regulators, flowmeters, and connecting hoses.

Plasma arc welding uses two gases, one as the welding medium, the other for shielding. The torch has a tungsten electrode and a small orifice. In this work, the argon gas is heated until it becomes partly ionized and able to conduct an electric current. A filler metal may be used. The secondary gas flows through a nozzle cup and surrounds the arc and the weld. This gas may be argon, helium or a mixture. Equipment is similar to that used for tungsten welding.

Resistance welding uses electrical current through electrodes to join overlapping metal sheets of different thicknesses. The sheets are clamped together tightly while the molten metal is applied at the joint.

Flash welding is usually done by machine and is almost always automatic. Heat at the joint is created by arcs and resistance to current flow and, when force is applied, there is usually a shower of sparks. This technique is usually applied in butt welding of rods and bars.

Electroslag welding uses a molten slag to heat the base metal edges and the filler and protect the weld. This is a

Figure 24-1. Flash welding operation. (American Welding Society)

machine process in which an arc melts the slag and preheats the metal. When the slag becomes conductive, the arc is stopped because the slag resists the flow of electrical current and the heating is continued.

Thermocutting by arc or plasma methods are different from the welding methods. In arc cutting, current is supplied by standard welding power. Electrodes consist of graphite and carbon, coated with copper to improve capacity for carrying electrical current. The metal that is melted in the welding process is blown away by compressed air.

Plasma cutting is achieved by a high velocity hot jet, created by forcing inert gas and an arc through a small orifice. The metal is melted by arc energy and expanding gases move the molten metal from the cutting area. This process requires high voltages, special power supplies, and water-cooled torches. Mixtures of argon-hydrogen and nitrogen-hydrogen are used.

Oxy-fuel Gas

Acetylene is the principal gas used in this process, although other gases and mixtures have been tried. These include propane, propylene, natural gas-methane, hydrogen, and methyl acetylene-propadiene (stabilized), most of which have limitations in welding.

The basic oxy-fuel gas process utilizes a flame at high temperature to melt the workpiece metal and the filler metal, while the flame is adjusted to provide a neutral environment so that the molten metal will not be contaminated before it becomes solid.

Brazing is a method by which the filler metal is distributed through a close joint by capillary action. The base metal is heated below the fusion temperature and a filler metal having a melting temperature below 840 °F is flowed into the joint.

Braze welding is used on ferrous metals, copper and nickel alloys, and sometimes on metals that are not similar. For this, the filler metal is directed into a groove or fillet. The edges of the joint are heated to a dull red, and flux is applied as the filler from a bronze rod is applied.

Other uses of oxy-fuel gas include surfacing, or the means of applying a layer of filler metal on a base metal; forming, or shaping and bending by heating; annealing, hardening, and softening with a flame; and removing rust and scale from metals by flame priming.

Scale can be removed from metals by flame in preparation for inspection or machining, and a torch can be applied for burning paint, finishing glass, antiquing wood and other practical work.

In oxy-fuel gas cutting, the fuel gas is blended with the oxygen in a high temperature flame to groove or cut metal. In a small operation, equipment might only include a cylinder of oxygen and one of fuel gas, a pressure regulator on each, the required hoses, and a torch. In a large plant, the oxygen might be supplied from a generating plant, from storage tanks of liquid oxygen or high pressure gas, or from manifolds. The thermal cutting might be accomplished by machine.

Another process is thermal spraying in which metal or nonmetal particulates are applied in molten condition to form coatings. This can be done by flame or plasma spray, or by electric arc spraying.

HAZARD CONTROL

Combustible Solids

Any unnecessary material should be removed from the area of cutting and welding, particularly combustibles. As an inspector, you should watch for patches of oil, grease, or other contaminants which might be exposed to the oxygen. It is well known that materials that burn normally in air will burn violently or explosively in oxygen under pressure. Cloths, papers, and smoking materials should not be in the work area. The cylinders, torch, and other equipment should be well cleaned before use.

Flammable Liquids

Welding and cutting operations should not take place in the vicinity of open flammable liquids, unless provisions have been made to prevent ignition from any source. Complete shielding of the torch operation may be necessary and this would require appropriate fire resistive materials and complete protection of the source of flammable liquid vapors.

Frequently, a torch is applied to a metal drum or other container that formerly held flammable liquid. This is a hazardous operation that has resulted in many injuries and some fatalities. Before such an operation is begun, the interior of the container must be completely purged of vapors and residues.

Spark Control

Sparks from the torch operations can be controlled to some extent by the torch user, but if work cannot be moved into a safe location, then shields or partitions may be used to control the direction of spark flight. In addition, one or more persons should be assigned to serve as fire watchers and be equipped with extinguishers or charged hose lines. After the torch work is completed the area should be checked for signs of burning or hot metal residue.

Fire Protection

Welding and thermal cutting may be used in almost any building or area, but as inspector you must judge the potential effectiveness of fire protection that exists at the site of operations. In a mill or industrial plant, there may be appropriate automatic and portable extinguishing equipment, but other buildings may not be so equipped. In your inspection you should verify that minimum practical measures at least are being applied. The torch operator should be adequately protected and informed of the fire and explosion hazards of the task. The supervisor or other representative of management should also be aware of the hazards and the need for safe procedures.

Appropriate portable fire extinguishers, hose lines, and fire resistive tarpaulins, shields and other protection should be available and in good condition, if the risks warrant such protection.

Floors should be cleared of combustibles and, if they are metal floors, they should be kept wet or be otherwise protected.

All personnel working in the area must be trained and informed about what to do if fire or explosion occurs. They should also be trained in first aid for flesh burns.

Personnel Protection

It is essential in welding and thermal cutting to give protection to the torch user, the helper, and other nearby workers. Their protective clothing must be adequate to resist the intense heat of molten metal sparks and slags, and they may need eye protection even if they are not personally involved in the work.

Appropriate protection includes flame resistant gloves of leather or other standard work material, woolen clothing, aprons of flame resistant material, cape sleeves or shoulder covers with skull caps under helmets, or with goggles for overhead work, leggings, and high top safety shoes. The goggles and helmet facepiece must be effective against the glare of molten flame and sparks.

Sleeves and collars should be kept buttoned to prevent spark entry. Clothing should have no front pockets. Trousers should not be cuffed or turned up on the outside. There

Figure 24-2. Attention to clothing is an important factor in operator safety. Note the protective mask and gloves. (Linde Div., Union Carbide Corporation)

should be no oil or grease on the outside of clothing. Cotton clothing may be worn if it is suitably treated to reduce its combustibility.

As an important part of your inspection, make sure that the welding and cutting personnel realize that oxygen should never be used to cool the welder or helpers, ventilate a confined space, or dust off clothing. Intense flame can result from these well-intentioned mistakes.

FUEL GAS SYSTEMS
GOOD PRACTICE CHECKLIST

Regulators

There should be no evidence of continuous pressure increase ("creeping"). Gage hand should return to pin on pressure release. Connection between regulator and cylinder should be gas tight.

Torches

Orifices in tips clean. Packing tight around torch stem. Valves shut off completely. Passages clear. No backfires or flashbacks (these usually caused by loose connections or improper seating). Torches free of oil and grease. Friction lighter provided.

Hose

Hose and connections made specifically for oxyacetylene apparatus. Oxygen hose green; acetylene, red. Oxygen hose connections right-hand thread; acetylene left-hand. No oil, grease, white lead or other pipe fitting compounds used for making joints. No unnecessarily long lengths of hose. Inspect for leaks, burns, worn places, loose connections, and other defects.

Piping

Oxygen piping steel, wrought iron, brass or copper; acetylene piping steel or wrought iron only (acetylene pressure not to exceed 15 pounds per square inch). No cast-iron fittings. Joints in steel or wrought iron pipe welded or made up with threaded or flanged fittings; joints in brass or copper pipe welded, made up with threaded or flanged fittings, or if of socket type made with solder with melting point over 1,000°F. Piping to be gas tight at one and one-half times maximum working pressure.

All piping run as directly as possible, and inside piping rigidly secured. Oxygen pipes and fuel pipes identified by name of gas on pipes at junctions and points of distribution. Oxygen piping kept free of oil. Provision for draining low points in all piping. Underground pipe beneath frost line. Buried pipe and outdoor ferrous pipe covered with suitable corrosion resisting material.

Valves provided to shut off supply to any building in emergency. Signs locating and identifying section shutoff valves. Hydraulic flash arrester or check valve on acetylene piping at outlets.

Oxygen Cylinders

Properly identified. Cylinders and fittings free of oil or grease. Equipped with pressure reducing regulator before use. Cylinder valves respond to hand pressure. Cylinders upright when in use. Protection against contact with wires, sparks, and open flames.

Acetylene Cylinders

Properly identified. Cylinders upright when in use. Protection against contact with electric wires, sparks, and open flames. Fusible plugs in good condition. Absence of leaks around valve stem. Special T-wrench or key provided for opening or closing cylinder valve. Equipped with pressure reducing regulator before use.

Acetylene Manifolds

Only cylinders containing gas at approximately equal

pressures manifolded. Where manifolded cylinders aggregate over 2,000 cubic feet of capacity, location outside in a special building or room or separation by 50 feet is required.

Oxygen Manifolds

Manifolds not located in acetylene generator room or close to cylinders of combustible gases. Where over 6,000 cubic feet of capacity is manifolded, location outside or in separate special room or separation by 50 feet is required.

SI UNITS

The following conversion factors are given as a convenience in converting to SI units the English units used in this chapter.

$$1 \text{ ft} = 0.305 \text{ m}$$
$$1 \text{ psi} = 6.895 \text{ kPa}$$
$$\tfrac{5}{9} \, (°F - 32) = °C$$
$$1 \text{ cu ft} = 0.0283 \text{ m}^3$$

BIBLIOGRAPHY

McKinnon, G. P. ed., *Fire Protection Handbook*, 15th ed., National Fire Protection Association, Quincy, MA, 1981. Chapter 11 of Section 9 includes a good summary of cutting and welding operations and an explanation of the required safeguards.

McKinnon, G. P. ed., *Industrial Fire Hazards Handbook*, 1st ed., National Fire Protection Association, Quincy, MA, 1979. This handbook also covers welding and cutting, but with more extensive explanations of industrial procedures and equipment.

NFPA Codes, Standards, and Recommended Practices. (See the latest *NFPA Codes and Standards Catalog* for the availability of current editions of the following documents.)

NFPA 51B, *Standard for Fire Prevention in Use of Cutting and Welding Processes*. A standard that explains the fire protection recommendations for these operations, fire prevention precautions, and duties of fire watchers. It also covers special measures required for public demonstrations.

NFPA 306, *Standard for the Control of Gas Hazards on Vessels.* This standard deals with vessels that are to be repaired. It is particularly useful for those who inspect ships in port that are carrying or have carried combustible or flammable liquids, compressed gases, bulk chemicals, or other hazardous cargo.

NFPA 327, *Standard Procedures for Cleaning or Safeguarding Small Tanks and Containers.* Explains the hazards of these operations and the procedures for flushing or inerting these containers to minimize flammable vapors before using a torch.

Chapter 25

SPECIAL OCCUPANCY HAZARDS

Almost all, if not all, industrial or manufacturing plants are faced with fire hazards of one sort or another. Many of these hazards may be common to all, but some processes present unique hazards, or if they have similar hazards, those hazards may be more severe. Fire control, suppression, and extinguishment may be much more difficult and the consequences more disastrous. This chapter points out the hazards specific to special occupancies without details of the processes themselves.

ALUMINUM PROCESSING

In aluminum processing plants, the primary fire hazards are the equipment for producing or regulating the large amounts of electric power. Other major hazards are hydraulic systems, conveyors, the use of combustible oils in rolling or forming, electrode production, and dust explosion or fire.

Alumina, a mixture of aluminum and oxygen, is precipitated out of a solution of bauxite and caustic soda. The alumina is dried in oil or gas-fired kilns at 1,800°F. Combustion controls for this equipment require special maintenance efforts to keep them reliable from a safety standpoint.

The alumina is smelted, or reduced, to aluminum in electric furnaces, called pots, by a high-voltage, low-amperage direct current. Major transformers and rectifiers needed to provide the large amounts of current usually are located in a switchyard. They should be arranged so that there is sufficient separation between oil-filled equipment.

The control building should be of noncombustible construction, and cable runs, pass-throughs, and openings firestopped. Cable trays should have noncombustible covers.

Electrodes for the furnaces are made from a mixture of inert carbon and combustible pitch in vessels heated by steam or oil. Hot-oil heating presents a fire hazard. The storage and handling of solid pitch also present fire and explosion hazards, as would any dust. Anodes are hardened in pits into which an oil or gas flame is introduced. Combustible volatiles are given off, and fire in the exhaust duct is a serious hazard. The baking process should be carefully controlled to prevent fire.

The primary hazard in smelting is the failure of a pot lining which would permit a molten metal spill called a *tap out*. Utilities should be so arranged that they will not be damaged by the spill.

Large amounts of combustible oils are used in the rolling of finished aluminum. That used in cold rolling is similar to kerosine in ignition and burning characteristics, and proper fire protection measures should be employed.

The handling of heavy ingots and coils usually requires hydraulic equipment utilizing combustible oils. Adequate maintenance and protection is required.

Although aluminum is a metal, it will burn, especially in the form of a fine powder or dust. This hazard is present in the scalping or planing of rough ingots to give them a smooth surface. Shavings and fines can produce a fire or explosion in dust collecting systems. Exhaust systems and collectors should be designed and protected as outlined in NFPA 65, *Processing and Finishing of Aluminum*.

ASPHALT MIXING PLANTS

Fire hazards in asphalt mixing plants are generally related to the aggregate dryer, the drum mix dryer, the pollution control system, and asphalt binder storage.

Tanks containing asphalt, cutbacks, and emulsion usually are heated by transfer oil. Transfer oil lines should be checked for damage that might permit contamination with asphalt, thus reducing the flash point of the oil. If the heater is direct fired, it should be equipped with a prepurge unit and high temperature shut-off, and properly protected against low asphalt level. The greatest danger of fire and explosion occurs when the asphalt does not cover the heating tubes.

Probably the most vulnerable equipment is the aggregate dryer, which develops temperatures up to 1,600°F. The dryer is subject to constant vibration so examine fuel lines, gas valves, flame guards, pumps, and motors for loss connections or leaks that might lead to fire or explosion. It also is important to see that personnel are aware of the need and trained to purge the heating and exhaust system after every shutdown. Spontaneous combustion can occur when there has been insufficient purging or when leaky valves or connections have allowed combustible gases to be drawn into the dryer.

Hot mix plants, whether portable or permanent, require constant inspection and preventive maintenance. Check to see that the interlock system shuts down the burner and the flow of asphalt is reduced when the flow of aggregate is lessened or stopped. Otherwise, the drum may be overheated and the asphalt ignited. The interlock must also prevent flames being drawn into the exhaust system and igniting asphalt droplets.

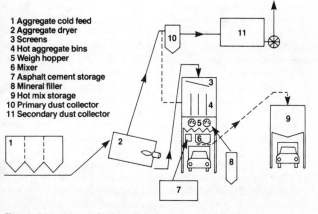

1 Aggregate cold feed
2 Aggregate dryer
3 Screens
4 Hot aggregate bins
5 Weigh hopper
6 Mixer
7 Asphalt cement storage
8 Mineral filler
9 Hot mix storage
10 Primary dust collector
11 Secondary dust collector

Figure 25-1. Materials flow diagram of an asphalt batch plant process.

Secondary collectors should be of the wet type to prevent spontaneous ignition of wood splinters or roots from bank-run gravel.

Fuel tanks usually are below ground, but above-ground lines, valves, and couplings should be protected by posts or stanchions against damage from truck traffic.

All general safety rules regarding welding and the use of solvents should be observed.

CLAY PRODUCTS

A large proportion of the fires in clay products plants are from common causes, such as high temperatures, storage and use of solid and liquid fuels, overheating due to insufficient clearance, and poor maintenance. The specialized hazards are small particle combustibles, straw and other combustibles for packing, welding, and careless smoking around combustibles made more ignitible by exposure to process heat.

Materials such as sawdust, ground nut shells or similar substances are sometimes added to the clay mix. Handling and storage of these present the explosion hazards of organic dusts. Damp sawdust may heat spontaneously and should be disposed of properly.

Napthalene, a combustible solid, is sometimes used in brick making. It sublimes when the clay is fired and may be recovered and reused. Napthalene vapors and dusts form explosive mixtures with air. Recovery equipment should be of safe design with adequate ventilation. Storage and handling of napthalene should be done in accordance with acceptable standards for volatile combustible solids.

Oil of kerosene class is often used to prevent clay from sticking to molds. Oil-soaked floors and benches add to fire spread. Safeguards against oil spills should be instituted.

The drying and firing of clay products require heat, often at high temperatures. The usual hazards of fuels, burners, and electrical and mechanical faults are present. Combustible materials must be safely separated from heat and ignition sources. Examine chimneys and waste heat ducts for cracks and holes.

All plants should have adequate fire extinguishing equipment for Class A fires, and employees should be trained in the proper use of fire fighting equipment.

COMPUTER CENTERS

Fire hazards in computer centers are generally associated with areas surrounding the computer room. Malfunctions within the equipment itself may cause components to overheat and produce smoke. These failures are readily detected and controlled with proper detection systems and portable fire extinguishers.

The major hazard comes from the paper forms, records, tapes, and boxes used in data processing and which should be kept in safe areas outside the computer room. Paper and other combustibles within the computer room should be only those absolutely necessary. These should be totally enclosed in metal cabinets.

Spaces beneath raised floors for cables or other purposes should be protected by smoke detectors and inspected for combustible debris.

Areas critical to life safety and computer operations, including air-conditioning and electrical equipment, should be provided with fire detectors and appropriate fire suppression equipment.

DRY CLEANING PLANTS

Special hazards of dry cleaning plants are determined to a large extent by the class of solvent used. Systems using Class I solvents, such as low flash point naphtha, are now prohibited. Type II systems use solvents with a flash point between 100°F and 140°F. Type III systems use solvents with flash points between 140°F and 200°F. Type IV and V systems use solvents considered to be noncombustible. They may, however, have toxic vapors and should be adequately ventilated.

The fire hazards stem from:

- Poor housekeeping, lint and trash accumulation, and lack of cleanliness.
- Improper wiring of equipment and overloading of circuits.
- Failure to properly maintain and operate equipment.
- Failure to maintain solvent temperatures at least 20 °F below flash points.
- Unsafe storage of chemicals and solvents.
- Static electricity, sparking from metal objects, and ignition of matches left in garments.

Type II and III plants should have scuppers, curbs, or special drainage systems to carry spilled or leaked solvents to a safe area. The exhaust system should have a capacity of 1 cubic foot per minute for each square foot of floor area. The amount of ventilation should keep solvent vapor concentration to 25 percent of its lower flammable limit.

Solvent storage in the cleaning room may not exceed two 1,500-gallon tanks.

Washing machines should be equipped with automatic solvent shutoff or overflow drains to an underground tank.

In Type IV plants, apparatus with open flames or exposed electric heating elements should be located so that cleaning vapors are not drawn over them and converted to phosgene, a toxic gas.

ELECTRIC POWER SOURCES

Large-scale electric generating plants may use fuel oil, coal, gas, water, or nuclear fuel to supply the necessary energy. Hydro (water) generating plants have no primary fuel fire hazards. The hazards of nuclear plants are covered in another section. Gas turbines are not used to produce the normal output of generating plants.

Hazards in coal- and oil-fired generating plants are associated with the storage and transmission of the fuel, boiler explosions, failure of lubricating and hydraulic oil lines, leakage of the hydrogen used to cool turbine generators, failure of oil-insulated transformers, and deterioration of cable insulation in spreader rooms.

Coal is subject to spontaneous heating in storage yards, bunkers, and conveying equipment. Such fires are difficult to control. All possible sources of ignition should be eliminated by prohibiting smoking and open flames. Accumulation of coal dust should be avoided by a continuously operating collection system and careful housekeeping.

Pulverized coal is entrained in air and carried to the boiler. Pulverizers should be equipped with separators to remove tramp metal that might cause ignition. Failure of the feed pipe, from erosion or other cause, can release an explosive cloud of dust. All equipment concerned with the storing, pulverizing and moving of coal should be equipped with automatic detection and suppression devices.

Oil-fired boilers often use heavy fuels, which must be heated for pumpability and atomization. All oil tank heaters should be equipped with thermal limiting devices to prevent overheating of the oil, and flow-through heaters should have a switch to stop the heating if circulation is interrupted.

All fuel-handling equipment should be located outside of tank dikes. Pumps should have filters to prevent damage, and a curb and drain system should be provided for pump seal leakage. Oil is carried to the boiler at about 135°F and at pressures up to 1,000 psi. Rupture of an oil line can cause an extensive spill or an atomized spray. Hot surfaces present an ignition hazard. Fuel oil piping should be of all-welded construction with a minimum of connections.

Boiler explosions result from uneven or low fuel flow causing a loss of flame in a portion of the boiler. Continued introduction of fuel builds an explosive "cloud" which often finds a source of ignition from other flames or hot boiler surfaces.

Turbine generators require substantial systems for lubricating and hydraulic oils, and these present significant hazards. Oil lines should be installed away from valve bonnets, steam pipes, steam chests, and other turbine parts operating at or above 700°F. High-pressure oil lines should be run inside another pipe to contain oil from leaks.

Tubine generators may be cooled with hydrogen which, if not carefully controlled, presents a severe hazard.

Furniture Manufacturing

Furniture manufacturing plants, except the most modern ones, present severe fire hazards. The buildings themselves are usually combustible. Wood, the primary raw material, is normally combustible and made more so by air-drying or kilning to 7 percent moisture. The process produces shavings and dusts and employs highly volatile and flammable solvents and finishes. High ventilation requirements produce large airflows that spread fire rapidly.

Despite the hazards presented by the materials and processes, however, fires are most likely to occur from poor maintenance and repair practices involving welding and similar activities, saw or grinder sparks, or overheated motors. Contaminated rags, accumulation of overspray, and poor ventilation leading to a build-up of solvent vapors are other contributing factors.

Poor housekeeping, smoking, boiler room sparks, and portable heaters or barrel fires for warmth are the main hazards in lumber storage yards. Storage piles should provide adequate aisles for fire apparatus.

Waste material from rough milling is carried by belt conveyor to the wood hog, which cuts the waste into small pieces. The conveyor should have a magnetic separator to keep tramp metal from causing fire in the hog. Waste material swept up and carried to the hog is a source of trouble because it often contains metal objects. Finer material is removed by air-moving equipment, cyclones, or bag filters. Some waste is used as fuel, and some may go to pulp or particle board manufacturers. Air-moving systems should be designed to remove transient flammable vapors.

GRAIN MILL PRODUCTS

The most serious hazard in the handling and storage of

grain is explosion of dust. By far the greatest number of dust explosions have been of undetermined cause or locations. However, the vertical bucket elevator is the most common location of primary dust explosions where it has been possible to determine the location.

Where it has been possible to determine the cause, welding or cutting torches, friction in bucket elevators, and fire, other than welding, rank closely as the most common ignition source.

Dryers also present a significant hazard in grain handling. Dryers are direct-fired with the heat of the burned fuel directed into a stream of air that is passed directly through the moist grain. Heated air leaving the dryer carries coarse dust and fine particles, and care must be taken to prevent their reentry into the burner.

The grain cleaning operation removes and concentrates extraneous materials from the grain. These materials, because of their dryness and high fiber content, are more prone to ignition than the grain itself.

Bucket elevators require constant care and maintenance to eliminate ignition sources. These can include bent or broken buckets, belt splice failures, electrical sparking resulting from overloading or grounding, open flames from smoking materials or space heaters. Mechanically heated surfaces, such as bearings, sheet metal in contact with moving belting, and vibrating dissimilar materials can generate enough heat to ignite dust. Vehicles with internal combustion engines may have surface temperatures high enough to ignite dust-laden air.

The all-important safeguard is the ventilating and dust collecting system.

LABORATORIES

Hazards in industrial laboratories can be classed as fire hazards and non-fire hazards. The latter are those which can cause injury without ignition or flame. They include radiation and caustic. irritating, and toxic chemicals. Some of these may become more dangerous under fire conditions.

Fire hazards may include compressed gas cylinders, electrically operated laboratory equipment, and static electricity. Combustible or flammable liquids or vapors, reactive chemicals, and flammable solids also are fire hazards, as are furnishings, equipment and paper products.

One way to reduce the fire hazard is to limit available fuel by excluding or protecting large glass containers of flammable chemicals, putting as much as possible in safety cabinets or rooms. The total amount of flammables should be limited to the quantity sufficient for convenient operation. NFPA 30, *Flammable and Combustible Liquids Code*, sets a limit of 25 gallons of Class IA liquids and 120 gallons of Class IB through IIIA liquids. NFPA 45, *Standard on Fire Protection for Laboratories Using Chemicals*, sets limits based on construction and arrangement of laboratory space.

Ignition sources should be limited by prohibiting smoking and frequently inspecting motors, switches, ovens, hot plates, and gas burners. Flammable chemicals should be stored in suitable compartments. Refrigerated compartments are considered hazardous locations if used to store flammable liquids that cannot be cooled below their flash points in ordinary refrigerators or freezers. Ethyl ether, pentane, and similar liquids should be stored in explosionproof or flammable materials refrigerators.

Laboratories should establish a system for strict control of welding, cutting, and other operations which might introduce ignition sources where combustible or flammable materials may be present.

MACHINE SHOPS

Metal machining may appear, at first sight, to have few, if any, fire hazards. However, there are many aspects of the operation that present serious fire hazards. One of these is that nearly all metals will burn in air under certain conditions, depending on size, shape, and quantity. Various metals react differently with liquids used as coolants or lubricants.

The metals most susceptible to ignition during milling or grinding are aluminum, magnesium, titanium, uranium, and zirconium. For most of these, dry extinguishants rated for Class D fires should be readily available and carefully applied.

The principal hazards to be found in metal-working or machine shops are:

- Chip fires at the machine caused by heat or friction.
- Spontaneous combustion of cuttings.
- Combustion of coolants and lubricants.
- Fine particles that are combustible or explosible.
- Explosible hydrogen evolved from reaction of metals such as uranium, aluminum, and magnesium with water.
- Escape of pressurized hydraulic fluids from machine tools and accessories.
- Combustion of oil vapors deposited on the building surfaces.
- Combustion of oil-saturated floors.

It often is necessary to clean or degrease machined pieces. This is often done with flammable solvents and constitutes a fire hazard. Trichloroethylene, a nonflammable solvent, will react with aluminum to form soft, burning carbon and great volumes of toxic and corrosive hydrochloric acid vapor.

Cuttings and chips should be collected in noncombustible containers and stored in safe locations. Cutting fluid spills and drippage should be soaked up and not allowed to accumulate. Building surfaces should be inspected frequently. Dust and oily residue should be cleaned as necessary.

MOTOR VEHICLE ASSEMBLY

Two categories of materials present most of the fire hazards associated with the assembly of motor vehicles. These are plastics and flammable liquids. Rubber tires and polyurethane-foam seats are particularly difficult to protect in storage, and fire involving them is likely to be extensive and serious.

The flammable and combustible liquids include gasoline,

lubricants, engine coolants, hydraulic fluid, paint, thinners, adhesives, cleaning solvents, and sealers. High volume liquids such as gasoline, coolants, brake and hydraulic fluids, and paint often are pumped from remote storage to point of use. Small quantities of flammable or combustible liquids may be kept in safety cans at work stations.

Probably the most common fire cause in finished vehicles is electrical. Mistakes in wiring assembly may result in crossed circuits and overheating of the wires. Metal fastening clips may cut through insulation and cause a short-circuit fire. Another fire source is carburetor backfiring when engines are started with air cleaners removed and carburetors primed with gasoline.

Welding is fairly common as a cause of fire in motor vehicle assembly. Large components are welded in machines using hydraulic clamps and pressure. Oil leaks are common and collected in a sump which may have other combustible debris. The welding arc is an ever-present ignition source. Cardboard cartons, wooden crates, and burlap bags used as containers for small parts are often brought to the welding area where they present a fire hazard. Body seams are sealed during and after welding. The sealant presents no special fire hazard, but solvents are required for cleaning up sealant spills. These are usually flammable and all solvent residue must be removed before welding is resumed.

Maintenance and housekeeping are extremely important. Areas susceptible to fire include exhaust systems, electrical equipment, and paint overspray collectors. Rubber dust ground off on the test rolls is subject to spontaneous ignition.

NUCLEAR ENERGY PLANTS

The fire hazards of nuclear power plants are low compared to those of fossil-fuel power plants. However, the consequences of any fire may be much more serious because of radiation.

The hazards are:

- Oil fires involving reactor coolant pump motors and

emergency turbine-driven feedwater pumps.
- Fuel fires at diesel driven pumps.
- Fires involving charcoal in filter plenums.
- Fires in cable insulation.
- Fires in combustible waste and organic resins.
- Fires of flammable off-gases.
- Fires in protective coatings.
- Fires of turbine lubricating oil and hydrogen seal oil.
- Leakage of hydrogen cooling gases.
- Leakage of hydraulic fluids from turbine controls.

Most of these hazards are identical to those found in conventionally fueled generating plants and the safeguards are the same.

Hydrogen off gas is flammable and radioactive. It is one of the most severe hazards and requires constant attention. It should be held in decay tanks prior to recombination with oxygen or venting to the atmosphere. Constant monitoring of off-gas systems for oxygen is a necessary safeguard.

Records indicate that the construction stage is the most vulnerable period for fire damage in nuclear plants.

PAINTS AND COATINGS

The production of nitrocellulose lacquer and aerosol charging of containers present the most serious fire hazards in the making of paints and coatings. The use of Class I solvents (flash points below 100°F) and varnish cooking are the next level of fire hazard, followed by the use of Class II solvents. Making waterborne coatings presents little hazard.

Accidental ignition of flammable vapors must be guarded against as must the uncontrolled release of flammable or combustible liquids.

The most common cause of ignition is static sparks. Equipment must be bonded or grounded to prevent the generation of static electricity.

Reactors are used in the manufacture of resins, latex, and other special products. The reactions can become exothermic and the reactors must be designed to control the reaction.

Nitrocellulose presents special hazards in handling and storage whether alcohol-wet, water-wet, or in chips. Nitrocellulose should be stored under cover in a cool place away from the plant. It is easily ignited by friction and drums should never be skidded on the floor. All spilled residue should be swept up immediately with a natural bristle broom, deposited in a covered metal container and wet with water. The container should be emptied each day and the nitrocellulose burned after it has dried.

Areas that process nitrocellulose should be periodically hosed down or cleaned by an explosionproof vacuum cleaner with a water chamber. Solidified residue on floors and equipment is highly combustible and difficult to extinguish.

Tools used around nitrocellulose and flammable liquids should be nonsparking. Covers on mixing tanks also should be of nonferrous metal.

PAPER PRODUCTS

Paper products are items such as cardboard boxes, milk cartons and a host of other products which start as rolled paper. The paper passes through machines which cut, shape and convert the wet stock into finished goods. Regardless of the end product, the fire hazards are similar and involve large quantities of combustible stock and combustible waste in the form of paper dust and scrap. Idle wooden pallets also present a significant fire hazard.

Stock for shipping containers comes in rolls about 87 inches wide. These generally are stored three high to a height of about 20 feet. High-piled rolled paper lends itself to rapid fire spread through the flue spaces between rolls. The fire spread is increased if the rolls are not prevented from unwinding to expose more unburned fuel. There is a trend away from high-piling stock by providing storage areas of larger square footage. Finished box blanks may be stored on rack storage shelves or on pallets piled one on top of another to the height of stability. Whatever the storage method, the hazard is directly related to the quantity of combustible material.

Most modern container plants have pneumatic systems for conveying waste to the balers. In some plants, waste goes to the balers in bulk carts from localized collection points. Corrugated cardboard tends to bridge or hang up in waste accumulators, creating an additional fire hazard. Baled waste becomes mushy and difficult to handle when wetted, presenting the danger of collapse and structural damage.

Printing operations in container plants present the usual hazards of flammable inks, solvents, driers, and cleaners. Similar hazards are found in other paper products plants.

PLASTICS FABRICATION PLANTS

Practically all plastics are combustible compounds which will burn under favorable conditions. Aside fom the basic combustibility of plastics, the conversion of feedstock into finished articles involves combustible dusts, flammable solvents, and hydraulic fluids, along with the storage and handling of combustible raw material and finished goods.

Fabrication plants are subject to a variety of hazards that can result in fire or explosion. These include those mentioned above and high heat elements, heat transfer fluids, static electricity, and poor storage and housekeeping practices.

Nearly all plastics will burn rapidly in the form of dusts. If dispersed in air they can be explosively ignited by spark, flame or metal surfaces heated above 700°F. Dust explosions are possible whenever plastics are pulverized, ground, machined, sanded, or compounded with dyes, fillers, lubricants, stabilizers, or modifiers. The basic chemical structure of the resin governs its explosibility. The addition of wood flour, cotton flock, or other combustible fillers usually increases explosibility.

Flammable solvents are used in nearly all plastic fabrication. They require the same careful storage and handling as they would in any other processing plants. They may be somewhat more hazardous when applied to plastics, as plastics generally generate and retain static electricity more readily than paper or cotton fabric. Static electricity can rapidly build up to spark discharge, a hazardous condition if dust or flammable vapors are present. Plastics do not have their static charge dissipated by high ambient humidity; thus, it is necessary to ground equipment and to be sure that tinsel conductors maintain contact with moving films.

Hydraulic systems used in plastics fabrication present the same hazards as they do in other industrial processes.

Storage arrangements for many plastics can be similar to those for any Class III commodity, such as wood, paper, and natural fibercloth. Polyurethane, polyethylene, plasticized polyvinylchloride, and polyesters present severe fire hazards, exceeded only by thermoplastics, such as polystyrene and acrilonitrile-butadiene-styrene (ABS). These materials break down, acting and burning like flammable liquids. As foamed material, these plastics present the most severe fire hazard.

PRINTING AND PUBLISHING

Hazards in the printing and publishing industry are due to the fact that the basic raw material, paper, is readily combustible. The oils and solvents used for printing and clean-up also are combustible and flammable, and the movement of paper through presses, folders and collators generates static electricity which can be an ignition source.

Letterpress operations produce an ink mist, which accumulates on building and equipment surfaces. Because of its oil content, the mist presents a fire hazard.

Offset printing may require oil-fired, gas-fired or electric heaters to obtain the proper humidity in the paper. Isopropyl alcohol, which has a flash point of 53°F, is used as a wetting agent in the water fountains of offset presses. Sheet-fed offset is dried with a cornstarch-based powder, which is more explosible than coal dust. This must be periodically vacuumed up and the cleaner bags emptied immediately.

Practically all printing involves frequent wash-ups and cleaning with highly volatile, flammable solvents, which may include lead-free gasoline from the local gas station. Silk-screen presses may be connected to drying ovens, creating the possibility of flammable vapor fires.

All solvents should be stored in safety cans, and data sheets should be checked for flash points. Water-type extinguishers should not be permitted in press rooms because of their incompatibility with flammable liquids and electrical equipment.

Storage of rolled stock presents serious fire hazards due to the flue-like spaces between rolls stored on end to heights of 20 feet or more. Storing rolls on their sides or on horizontal rods does not greatly alter the fire characteristics. Adequate sprinkler protection is a requisite for roll paper storage. Paper in sheet form on skids or pallets presents no unusual fire problem. Paper dust should not be allowed to accumulate on top of rolls or on building surfaces.

Good housekeeping is necessary throughout the plant to remove paper scraps, ink mist, or drying powder accumulations. Electrical equipment should be well maintained and grounded, and suitable arrangements made for grounding static electricity.

Smoking, the use of open flames, and hot surfaces should be carefully controlled.

PULP AND PAPER MILLS

Many areas of pulp and paper mills present severe fire hazards, primarily because of the combustible nature of the raw material and finished products.

Log piles may be either ranked (like cordwood) or stacked (dumped at random in cone-shaped piles). Dried wood and refuse combined with uncontrolled ignition sources such as welding, smoking, open burning, or poorly maintained handling equipment can cause a serious, difficult to control, fire.

Debarkers and chippers produce large quantities of readily combustible material. Chief sources of ignition are careless welding, friction, and electrical defects.

Major hazards in the bleaching process involve the storage and handling of flammable liquids or toxic gases such as methanol and chlorine. Electrical cables required for power and control are subject to corrosive atmospheres and chemical spills.

The boiler in which process chemicals are recovered can be subjected to a violent reaction if water comes in contact with the hot smelt. Fire can also occur in combustible residues in the flue gas circuit.

Boilers for meeting the steam requirements of the mill may be fired by gas, oil, coal, bark, or a combination of these, and are subject to the hazards of similar boilers in other industrial plants. The lime kiln can be oil- or gas-fired and presents fuel-explosion hazards similar to the power boiler. Noncondensible gases, such as hydrogen sulfide, originating in various processes are collected and piped to the kiln for incineration.

Tubrogenerators are equipped with central lubricating and hydraulic systems. Oil leaks can present serious fire hazards.

The paper machine is a major fire hazard. Paper scraps, wool, or synthetic felt, lint, and oily deposits can accumulate on the machine, hood, and duct surfaces. A break in the paper web can send large quantities of paper into the broke pit below the machine. Normally this is conveyed back for repulping, but it does present an added fire hazard. Dryers and coaters require heat that can ignite the paper if not carefully controlled. Flammable or combustible solvents may be used to clean calender rolls and other equipment. Extrusion machines combine the hazards of paper, polyethylene, and lint with electrically heated dies and gas-fired flame impingement heaters.

RUBBER PRODUCTS

The principal hazards in the manufacture of rubber products involve storage of natural and synthetic rubber and such compounds as sulfur, oil, and hydrocarbon solvents; mixing of solvents and rubber; mixing rubber; spreading cement; spreading rubber on fabric; and dipping rubber fabric in cement. The process involves a continuity of combustibles.

Heat created in the mixing operation, static electricity, buffing, spontaneous combustion, and such common operations as welding and burning are the chief sources of ignition.

Baled raw rubber is usually shipped and stored in wood or cardboard containers. Palletized and stored to a height of 12 to 14 feet, these materials present a significant fire hazard.

Rubber is broken up and compounded with carbon black and oil in a banbury mixer. Friction within the mill heats the compound to a predetermined level. Malfunction of the temperature indicator may allow the mixture to overheat and vaporize the oils to the auto ignition point. On exposure to air, the vapors ignite. Fire in the banbury can easily spread to the exhaust duct and dust collector. Periodic cleaning of the mixer is necessary to remove the oil-soaked residue which accumulates on the outside.

To improve adherence of rubber components, they are dipped or spread with a cement made of rubber, flammable liquids, and other compounds. The opening of the mixer may release flammable liquids or vapors. Ignited by static electricity or friction sparks, the ensuing fire can involve the whole room. The provision of 50 percent or greater humidity helps to control static electricity.

If dipping in rubber cement containing a low flash point flammable solvent is required, it is usually followed by drying in a heated oven. Adequate ventilation is necessary to keep the concentration of vapors under 25 percent of the lower explosive limit (LEL).

Daily welding and cutting is a usual practice. This ignition source needs careful control because of the constant presence of flammable liquids, combustible oils, and rubber materials.

SHIPYARDS

Shipyards are threatened by many fire hazards which arise from the proximity of noncompatible processes such as welding and painting. Open-flame work is the prime hazard. All welding and cutting must be carefully monitored because of this hazard.

The materials used in the construction of ways determine the fire hazards. When the ways are of wood, exposure to open-flame and electric-arc welding is of primary concern. No storage of materials or occupied structures should be permitted below the vessel being constructed. Accumulated flammable trash on or below the ways or aboard the vessel can lead to a catastrophic fire. Ways should be protected with adequate portable extinguishers, standpipes, and hoses. Substructure automatic sprinklers and vertical draft stops are desirable.

Cylinders supplying welding torches should be restrained in an upright position, and lines should be protected against injury and leakage. Electrical cables should also be protected and generators secured. Generators, gas manifolds, and cylinders should be located on the open deck or overside of the vessel.

Paints, solvents, adhesives, lumber, and other combustible materials should be limited aboard the vessel or closely adjacent to it. Crating and flammable packing materials should be immediately removed from the area.

When a vessel is ready for launching, all flame-producing work should cease; all flammable and volatile materials removed; and the ways properly greased. After launch, skid grease should be immediately removed.

The fire hazards of shipyards will vary with the size of the yard and the types of materials used for construction. As mentioned, the major hazards arise from open-flame operations and the presence of large quantities of fuels such as flammable gases, liquids, solvents, adhesives, paints, and wood. All operations should be carefully monitored to see that they are compatible. Smoking should be confined to safe areas and a high standard of housekeeping maintained.

TEXTILE MANUFACTURING

Production processes in the textile industry differ because of the raw materials used. However, they share certain fire hazards, such as storage and opening of bales, cleaning raw stock, weaving, and finishing. The extent of the hazards will vary according to the inherent combustibility of the raw material.

Cotton fibers are easily ignited by sparks or open flame and burn readily. Synthetic fibers are less easily ignitible, but burn readily. Wool and other raw materials may not exhibit the same ignitibility or burn as easily, but collectively they present the same fire problems as cotton.

Cotton and other natural fibers are shipped and stored in rather loosely packed bales, often covered with burlap. They are often stacked up to 20 feet high. Fire from a smoldering or "firepacked" bale or from another source can flash over to the surfaces of other bales and involve the entire stock.

Fires in the opener, or feeder, room can be caused by friction sparks from unremoved pieces of bale ties or from layering in the beaters. Lint accumulations on floors and machines are ready fuel for fires.

Card machines further clean and straighten the fibers. Fires in such machines are usually the result of friction sparks from tramp metal. Automated machines combining opening, picking, and carding are less susceptible to fire because of incorporated firesafety measures. However, fire in such equipment will spread through ducts to other machines or to the duct collector. Quick detection and suppression are necessary to avoid extensive damage.

In the combing, drawing, roving and spinning operations, there is danger of fire as the result of the bunching of fibers and ignition by friction.

Starch is used in the sizing of warp or filler yarns. Dust explosion can occur if the dry starch is improperly handled.

Fires in weaving rooms account for almost half of the fires in textile plants. The main cause of such fires is electrical. Vibration at the loom causes breaks in the wiring, resulting in short circuits and sparking. Lint and dust accumulations, unless periodically removed, provide fuel for such fires.

The manufacture of carpeting requires heat for setting the latex backing and setting the carpet width. Malfunction of gas-fired equipment, lint build-up, and electrical arcing are the most frequent causes of fire in carpet ovens.

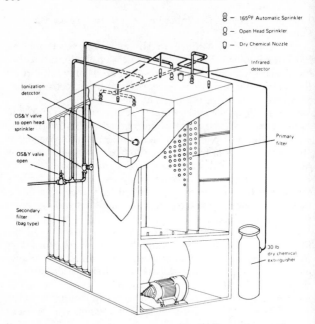

Figure 25-2. Arrangement of fire protection for a textile card filter unit. In this installation, there are two basic extinguishing systems — water and dry chemical.

VEGETABLE AND ANIMAL OIL PROCESSING

Fire hazards of processing vegetable and animal oils primarily involve the equipment and materials used as adjuncts to the main processing. Examples are the generation and use of hydrogen to harden fats and oils, the use of high temperatures and pressures and the spray drying of soaps and detergents.

Fats and oils tend to solidify in storage and must be heated before processing. Heating systems for storage tanks must enter at the top and extend to the bottom to provide a channel for the heated material to expand upward. Explosions can occur when tanks are improperly heated.

Disposal of the spent filter cake used in refining, degumming, and bleaching of oils and fats presents the hazard of spontaneous combustion. It should not be stored in the plant, but removed and buried. If flammable solvent is used to extract oil from the spent earth all the required safeguards for the use and storage of such solvents must be observed and the spent earth treated accordingly.

Because hydrogen has an extremely wide flammable range, there are fire hazards in the hydrogenation process. All electrical equipment should be suitable for Class I, Division 1 hazardous locations. Tools should be nonsparking, and employees should wear rubber-soled shoes.

The manufacture and storage of hydrogen require that all of the precautions for the hydrogenation process, and more, be taken. It is necessary to rely on the expertise of the plant designers and suppliers for full protection against fire and explosion.

Deodorizing requires temperatures of 210 and 274°C. Boilers for heat transfer systems employing the most frequently used mixture of diphenyl and diphenyl oxide are direct-fired with gas or oil. Leaks between firebox and boiler can result in fire or explosion. Because of low surface tensions and viscosities of heat-transfer media, welded construction is recommended.

The manufacture of soap, in addition to the above hazards, has two more. In the spray drying of soap to make beads, fire can occur in the spray tower or exhaust air ducts. Residue may stick to the tower walls and superheat to the point of ignition. Periodic cleanout can minimize this hazard.

Leaks in the fatty acids still and fractionating tower can permit the fatty acids to soak into the insulation. There, due to auto-oxidation, they may build up temperatures to the point of self-ignition.

WOOD PRODUCTS MANUFACTURING

Wood products plants include sawmills, which produce

standard-sized lumber, and plywood and particle board plants. Because they use raw materials and produce wastes similar to those of the pulp and paper and furniture industries, their fire hazards are similar in many respects.

Sawmill logs are stored in ranked or stacked piles or floated in ponds. Fire hazards include ignition from grass, brush, or forest fires, sparks from refuse burners, boiler stacks, locomotives, and vehicle exhaust. Waste materials, sawdust, bark, and scraps are conveyed to wood hoggers. Tramp metal entering the hogger can cause sparking and ignition of the waste. Combustible hydraulic oil in process machinery, temporary heating devices, welding sparks, faulty electrical equipment, and smoking all represent fire hazards. Good housekeeping is essential to reduce the amount of fuel available.

Stickered lumber in kilns and storage yards presents a severe hazard because of the horizontal flues which increase the burning rate and make extinguishment difficult. Planing of seasoned lumber creates dusts and fine shavings.

Particleboard and hardboard are made of sawdust, shavings, and chips. Dumping and retrieval of the raw materials can put wood dust into suspension and create an explosion hazard. Grinding raw materials to the desired size presents an explosion hazard from tramp metal. Drying of the ground material, whether in direct-fired or steam heated dryers, is basically hazardous.

Hardboard and fiberboard are made from wood fiber which is puffed in a steam chamber. Particleboard is made of wood chips. The forming and pressing operations required for each present explosion or fire hazards. After pressing, the boards are humidified in ovens where there, again, is a fire hazard. The dust collecting systems and large belt sanders used in the finishing operation are particularly susceptible to fire and explosion.

Plywood plants do not have the same level of hazards as do particleboard plants. They do have similar dust hazards in finishing operations. Equipment such as hot presses, sanders, and dryers are potential ignition sources.

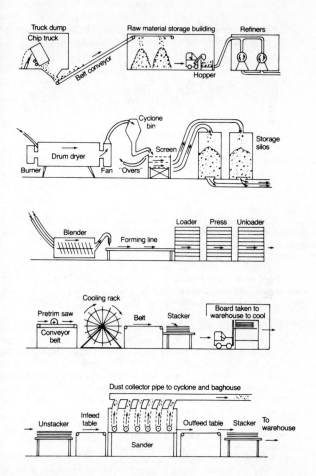

Figure 25-3. A flow diagram of a basic particleboard production process.

SI Units

The following conversion factors are given as a convenience in converting to SI units the English units used in this chapter.

$$1 \text{ in.} = 25.4 \text{mm}$$
$$1 \text{ ft} = 0.305 \text{ m}$$
$$\text{\%} \ (°F - 32) = °C$$
$$1 \text{ gal} = 3.785 \text{ litres}$$
$$1 \text{ cu ft} = 0.0283 \text{ m}^3$$

BIBLIOGRAPHY

McKinnon, G. P. ed., *Industrial Fire Hazards Handbook*, 1st ed., National Fire Protection Association, Quincy, MA, 1979. Provides a comprehensive description of the risks and hazards in the industrial environments discussed in this chapter.

McKinnon, G.P. ed., *Fire Protection Handbook*, 15th ed., National Fire Protection Association, Quincy, MA, 1981. Section 8 of this handbook discusses in detail many of the industrial occupancies covered in this chapter.

NFPA Codes, Standards, and Recommended Practices. (See the latest NFPA *Codes and Standards Catalog* for the availability of current editions of the following documents.)

NFPA 30, *Flammable and Combustible Liquids Code*. Discusses requirements for tank storage, piping, valves and fittings, container storage, industrial plants, bulk plants, service stations, and processing plants for flammable liquids.

NFPA 33, *Standard for Spray Application Using Flammable and Combustible Materials*. Addresses the application of flammable or combustible materials as a spray by compressed air, hydraulic atomization, steam, or electrostatic methods.

NFPA 45, *Standard on Fire Protection for Laboratories Using Chemicals*. Establishes basic requirements for the protection of life and property in laboratories handling hazardous chemicals.

NFPA 65, *Standard for the Processing and Finishing of Aluminum*. Covers fire protection requirements for industrial operations in which fine metallic aluminum dust or powder is liberated.

NFPA 66, *Standard for Pneumatic Conveying Systems for Handling Feed, Grain, and Other Agricultural Dusts*. Provides guidance on design safeguards of component parts of pressure type and suction type pneumatic conveying equipment.

NFPA 70, *National Electrical Code*. Specifies wiring and equipment requirements in domestic and industrial occupancies.

NFPA 75, *Standard for the Protection of Electronic Computer/Data Processing Equipment*. Contains fire protection requirements for data processing installations needing fire protection or special construction.

NFPA 90A, *Standard for the Installation of Air Conditioning and Ventilating Systems*. A standard to restrict spread of smoke, heat, and fire through duct systems; minimize ignition sources; and permit the use of duct systems for smoke control purposes.

NFPA 91, *Standard for the Installation of Blower and Exhaust Systems for Dust, Stock, and Vapor Removal or Conveying*. Provides the requirements for fans, ducts, duct clearances, system design, and dust collecting systems.

Chapter 26

SIGNALING SYSTEMS

Signaling systems perform several functions vital to limiting life and property losses during fires. They provide detection, early warning, and fire department notification.

PROTECTIVE SIGNALING SYSTEMS

Though there are five types of protective signaling systems, there are some basic features common to all systems regardless of type. Each has a primary power supply, and most have a secondary supply to operate the system in the event of failure of the primary supply. The systems have signal initiating circuits which are driven by automatic fire detectors, manual pull stations, waterflow alarm devices, and other initiating devices. They are also equipped with one or more indicating device circuits that operate audible and visual alarm notification devices, such as bells, horns, loudspeakers, lights, punch registers, and digital displays.

Protective signaling systems are usually tested and inspected on a regular basis by an alarm system service contractor or central station system owner. During your inspections, you should at least look at the signaling system devices located on the premises and report any damage or defects that you see. Note the location and accessibility of alarm boxes and the condition of exposed wiring. You should not make any operating tests of your own without the permission of the party responsible for the system. When you are going to conduct a test notify in advance all persons who would automatically receive an alarm so that a false alarm will not be transmitted.

Local System

A local protective signaling system is intended only to

sound a local evacuation signal in the protected building. It does not automatically relay the alarm to the fire department; someone must notify the fire department, by means of either the telephone or a municipal fire alarm box. Local systems are usually owned by the building owner or occupant. They are not required to be equipped with secondary power supplies.

Auxiliary System

An auxiliary system is basically a local system that has provisions for connecting to the municipal fire alarm system through a master box. This system does transmit an alarm to the fire department. The alarm received by the fire department would be the same as that received if someone had tripped the municipal fire alarm box, but department records would indicate that an auxiliary system was connected to that box. Auxiliary systems must have a secondary or standby power supply. Although the system up to the point of connection with the municipal fire alarm box is usually owned by the building owner or occupant, it may be leased from an outside service, which would be responsible for maintenance and testing.

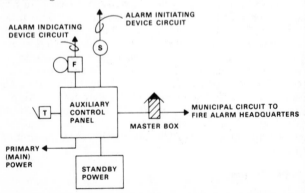

Figure 26-1. A typical auxiliary fire alarm system.

Remote Station System

This system is similar to an auxiliary system, except that it transmits an alarm to a remote location that is attended by trained personnel at all times. The remote receiving location may be a police or fire station or a telephone answering service. The signal is transmitted over leased telephone lines and is indicated both visually and audibly. System trouble signals are transmitted automatically to the remote receiving location. Remote station systems are also required to have a standby power supply. The owner or occupant of the property usually owns, or contracts for, the interior portion of the system and the equipment at the remote station, and leases the circuits between.

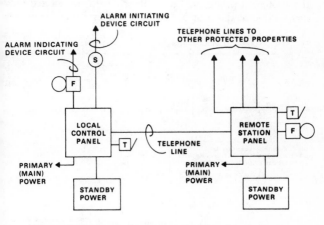

Figure 26-2. A typical remote station fire alarm system.

Proprietary System

A proprietary system transmits the alarm to a central supervisory station located in or near the protected property. The system is operated by someone who has a proprietary interest in the protected building, the owner or occupant for example. These systems are required to have a secondary power supply. In the past, most proprietary systems had separate initiating device circuits for each zone within a building. Modern systems, however, employ signal multiplexing and minicomputer systems that receive signals from throughout the building over a single pair of wires and determine the location of the fire.

Central Station System

These systems are similar to proprietary systems, except that the receiving station may be a considerable distance from the protected property. Each subscriber to the service may have a line to the central station, or several may be on the

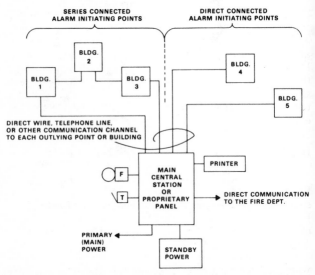

Figure 26-3. A typical proprietary or central station fire alarm system.

same circuit and each is given an identifying code number. Upon receipt of an alarm, central station personnel notify the fire department. Central station systems require standby power supplies.

SUPPLEMENTARY FUNCTIONS

Modern fire protective signaling systems can do much more than transmit an alarm of fire. They may operate smoke control equipment, control elevators, trigger special extinguishing systems, or control energy management systems.

Combination systems may include, in addition to fire alarm functions, burglar alarm, door entry control or paging systems.

FIRE DETECTION DEVICES

The most common elements of a fire that can be detected are heat, smoke (aerosol particulate), and light radiation. Detection devices should be inspected and tested according to the schedule specified for the kind of service and type of signaling system employed. You should review the records of tests made by those responsible for the maintenance of the equipment.

Heat Detectors

Heat detectors respond to convected thermal energy of a fire and are generally located on or near the ceiling. They respond when the detecting element reaches a predetermined fixed temperature or to a specified rate of temperature change.

Fixed Temperature: These detectors produce an alarm when the detecting element reaches a predetermined fixed temperature or when the temperature ranges at, or above, a specified rate.

One form of fixed temperature detector is the fusible element—eutectic metal alloys that melt rapidly at a predetermined temperature. This form is commonly used in automatic sprinklers and to release fire doors. Fusible elements are also used to actuate an electrical heat detector known as a spot-type detector. Following operation, either the operating element or the entire device has to be replaced.

An alternative to spot-type heat detection is continuous line detection, which may use replaceable fusible elements, a semiconductor material, or bimetallic elements. The bimetallic type are automatically self-restoring.

Rate Compensated: This type of detector responds when the temperature of the air surrounding the device reaches a predetermined level, regardless of the rate of temperature rise.

Rate of Rise: The rate-of-rise detector will function when the rate of temperature increase exceeds a predetermined value, typically around 12 to 15°F per minute. These detectors are designed to compensate for the normal changes in ambient temperature that can be expected under nonfire conditions. Rate-of-rise detectors are available in line or spot type.

Sealed Pneumatic Line: This unit consists of a capillary tube containing a special salt that is saturated with hydrogen gas. At normal temperatures, most of the hydrogen is held in the salt, and the pressure within the tube is low. As temperature increases at a point, hydrogen is released from the salt and pressure rises until a pressure switch operates, indicating that fire has been detected.

Combination Detectors: Some detectors may contain more than one element to respond to a fire. They may be designed to operate from the response of either element or from the response of both elements. An example is a detector that operates on both the rate-of-rise and fixed temperature principles.

Thermoelectric: The sensing element in this type of detector is a thermopile, which produces an increase in electrical potential (voltage) in response to an increase in temperature. The potential is monitored, and when it increases at an abnormal rate, an alarm is initiated.

Smoke Detectors

Smoke detectors are identified by their operating principle—ionization or photoelectric. Most fires are detected earlier by ionization detectors than by photoelectric detectors.

Ionization Detectors: These detectors employ a small radiation source which ionizes the air in a sensing chamber, causing a small current to flow through air between two charged electrodes. This establishes an electrical conductance in the chamber. Smoke particles entering the chamber decrease conductance. When conductance falls below a predetermined level, the detector responds.

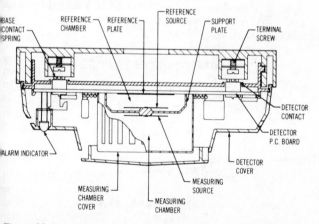

Figure 26-4. A cross-section view of an ionization smoke detector.

Photoelectric Detectors: The presence of smoke particles affects the propagation of a light beam passing through the air. Fire is detected by the obscuration of light intensity over the beam length or by scattering of the light beam.

Gas Sensing Detectors

Observations of large-scale fire tests have indicated that detectable levels of gases are reached after detectable smoke levels and before detectable heat levels. This has given rise to the development of semiconductor-type and catalytic element gas detectors.

Flame Detectors

Flame detectors respond to radiant energy visible to the human eye, or to radiant energy outside the range of human vision. These detectors are sensitive to glowing embers, coals, or flames. Because flame detectors must be able to "see" the fire, make sure that they are not blocked by objects placed in front of them. Infrared detectors must be shielded from the sun's rays, while ultraviolet detectors are insensitive to both sunlight and artificial light.

SUPERVISORY SERVICES

Actuating devices and signal transmitters used in supervisory services should be inspected and tested every six months or more frequently according to the schedule specified for the kind of signaling system employed. These inspections and tests are usually conducted by the service or contractor responsible for maintaining the system. However, you should review the records of those tests and inspections.

WATERFLOW ALARMS

Automatic sprinkler systems also serve as automatic fire

detection systems. They are equipped with devices to give an alarm when water flows, either a water motor gong or an electric switch. The switch is actuated by the alarm check valve, by a vane in the pipe, or, in the case of a dry-pipe system, by a pressure-responsive mechanism. Review any available records of system operation, and check wiring, gongs, and transformer or battery of local electric alarms. In cold weather, it may be impractical to run water through the water motor gong to test it. Waterflow alarms on wet-pipe systems can be tested by opening the inspector's test pipe at the top of the system. The alarm also can be tested by opening a bypass valve around the alarm valve on the dry-pipe valve, which tests both water motor and alarm switch. There is also a test switch in the circuit of a local electric bell. Where waterflow signals go to a central station or to the fire department, obtain permission and notify all parties concerned before conducting a waterflow test.

SI UNIT

The following conversion factor is given as a convenience in converting to an SI unit the English unit used in this chapter.

$$\%\,(\,^\circ F - 32) = \,^\circ C$$

BIBLIOGRAPHY

McKinnon, G.P. ed., *Fire Protection Handbook*, 15th ed, National Fire Protection Association, Quincy, MA, 1981. Section 15, Chapter 2 discusses signaling systems and their requirements, while Chapter 4 describes the various methods of fire detection.

NFPA Codes, Standards, and Recommended Practices. (See the latest *NFPA Codes and Standards Catalog* for the availability of current editions of the following documents.)

NFPA 71, *Standard for the Installation, Maintenance and Use of Central Station Signaling Systems*. The requirements for a system or group of systems maintained and supervised by trained personnel from an approved central station.

NFPA 72A, *Standard for the Installation, Maintenance, and Use of Local Protective Signaling Systems.* A standard on fire alarm or supervisory signals within the protected premises primarily for the protection of life.

NFPA 72B, *Standard for the Installation, Maintenance and Use of Auxiliary Protective Signaling Systems.* Signaling requirements for an individual occupancy or building or group of buildings, where the municipal fire alarm facilities are used to transmit an alarm to the fire department.

NFPA 72C, *Standard for the Installation, Maintenance and Use of Remote Station Protective Signaling Systems.* A standard for the direct connection of circuits between the protected premises and receiving equipment in a remote station.

NFPA 72D, *Standard for the Installation, Maintenance and Use of Proprietary Protective Signaling Systems.* A standard for systems under constant supervision of trained personnel.

NFPA 72E, *Standard on Automatic Fire Detectors.* Covers the minimum performance, location, mounting, testing and maintenance requirements of automatic fire detectors.

Chapter 27

WATER SUPPLIES

When you inspect an industrial plant, or any other manufacturing, storage, or processing building, you should give considerable attention to kind and amount of water supply available for fire protection. Within a community the water may come from the municipal source; outside the community it may be provided from a private source; in some locations it may be a combination of municipal and private supply; properties of state and federal agencies may have their own, supplemented by local supply.

Whatever the situation, the water used for fire protection usually is segregated by check valves from the potable water that otherwise serves the community or the industrial plant. This prevents possible contamination of the public supply. The private source of supply may be from a surface or ground source. The municipal system usually is supplied from a reservoir or other large body of water and features underground mains, pumps and hydrants, all designed and valved to maintain adequate flow for all normal needs.

For industrial plants and individual buildings the principal fire protection use of water is for supplying sprinkler systems, standpipe systems, and yard hydrants, and as an inspector you must become familiar with all of these, as well as the overall fire protection needs of the particular property.

A definition of a sprinkler system intended for fire protection is: an integrated system of underground and overhead piping designed in accordance with fire protection standards. It includes a water supply, an aboveground network of specially designed piping and sprinklers, a controlling valve, and a device for actuating an alarm when the system is in operation.

A standpipe system consists of piping, valves, hose outlets, and related equipment installed in a structure, with hose outlets for supplying handlines and nozzles. Many buildings have standpipe systems to supplement sprinkler protection.

Yard hydrants are those installed on the building's property with outlets appropriate for supplying pumping apparatus and hose lines. They may receive the water from the private or municipal source.

SYSTEM INSPECTION

As inspector you will need to know the components of each of these systems and their functions. You can begin with the water supply.

Maps and other information should be available within the plant or from an insurance rating bureau. You will need to identify the location on the property where water is received by the fire pump, then identify whether the water is of municipal or private supply. It will be important to record the pumping capacity and test performance, and whether the water supply is limited or unlimited. Standard inspection forms are available to facilitate your examination of the installations. These are prepared by the installing contractor, and copies are given to the building owner and local approving authorities. They include plans and descriptions of installations, tests, sprinklers, piping and valves and related comments.

Sprinkler Systems

There are five general classes of sprinkler systems: wet-pipe, dry-pipe, preaction, deluge, and combined dry-pipe and preaction. There are also special systems designed for unique problem situations.

A wet-pipe system is charged with water and discharges immediately when sprinklers are opened by fire. Pressure gages are in the sprinkler risers, above and below each check valve.

In a dry-pipe system the piping contains air or nitrogen under pressure and water does not flow until a sprinkler opens, the dry-pipe valve opens, and water enters the system.

This kind of system is used in areas where freezing temperatures occur.

A preaction system contains air that may or may not be under pressure and has a supplemental fire detection system in the same area as the sprinklers. When the detector is actuated, a valve opens and water flows to the sprinklers.

A deluge system functions in a similar manner, but has open sprinklers on the piping system. When the valve opens, water is discharged from all the sprinklers.

The combined dry-pipe and preaction system also retains air under pressure and is connected to a fire detection system. When a detector is actuated, tripping devices open dry-pipe valves simultaneously, without loss of air pressure in the system. Air exhaust valves at the end of the feed main open and water goes to the sprinklers. The fire detection system also signals a fire alarm.

When inspecting each of these systems, you will need to know the system layout, the locations and kinds of pressure and flow gages, the locations and functions of valves, the means of supplying air or nitrogen to a system, the required temperature for water, whether antifreeze solution is used, and the results of the most recent tests and flushing. (See Chapter 28 for more details.)

Standpipe Systems

Standpipe systems are intended for three general types of service: for use by fire departments and other personnel trained in the use of 2½-inch handlines; for use by building occupants who will operate 1½-inch handlines until the fire department arrives; or a combination of 2½- and 1½-inch hose to be used by the fire departments and building occupants.

These may be wet or dry systems, or a combined system that serves the standpipe and sprinkler system. They feature closets and cabinets for storing hose and nozzles.

Standpipes that do not exceed 100 feet in height must be at least 4 inches in diameter; those taller than 100 feet must be at least 6 inches in diameter. The maximum height allowed is 275 feet unless the building is zoned. The minimum flow required for normal systems is 500 gpm.

Standpipes that supply only 1½-inch hose must flow at least 100 gpm; those that supply 2½- and 1½-inch hose must flow at least 500 gpm. Appropriate hose outlets must be available on each floor. There should also be a fire department connection at the street side of the building.

Pressure gages are connected with each discharge pipe from the fire pump and public waterworks, at the pressure tank, at the air pump supplying pressure tank, and at the top of each standpipe. A water flow alarm may also be required.

All portions of the system must be inspected periodically, flushed and tested hydrostatically. Tanks must be filled properly, and at least 75 psi must be maintained in pressure tanks.

Look at the valves in the main connection to the automatic sources of water supply; they should be open at all times. Examine the hose, couplings and nozzles in the racks to verify that they are properly connected and have the right size diameter and threads.

Yard Hydrants

Yard hydrants must be of approved type with not less than a 6-inch connection to the mains and a valve in the connection. Outlets must have American National threads and may have independent gate valves. There may be exceptions to these requirements if local hose coupling threads do not meet the standards.

The number of yard hydrants required is determined by building and property. They must supply at least two hose streams for every interior part of the building not covered by standpipe connection and every exterior part by hose normally attached to hydrants. Hose lines should not exceed 500 feet in length.

Hydrants should be flushed and tested at least annually.

Inspection of Hydrants

(a) Is hydrant set up plumb with outlets approximately 18 in. above ground? Is the hydrant unobstructed and easily accessible; clear of snow in winter?

(b) Does hydrant drain properly? There is a small drain in the base of the barrel of frostproof types. When the main valve is open this drain is closed, but it is arranged to permit water to drain out of the barrel when the main valve is shut. If hydrant has been properly installed (with about a barrel of small stones under it), there is provision for water to drain away. If the drain is working properly and the main valve is tight, difficulty due to freezing of water in the barrel will be avoided.

(c) Open and close the hydrant to determine that it is working properly. Note direction of turn and number of turns to open fully. It is useful to post a sign on the hydrant showing number of turns to open and the direction of turn. The inspector can feel a suction at the outlets immediately after the valve is closed if the drain is working properly.

(d) Check for leaks. The main valve should close tight. There should be no flow from the drain valve when the main valve is open wide with the hydrant outlets capped. Look for leaks in mains near the hydrant. There are stethoscope-like listening devices which may be used for the purpose.

(e) Check freezing in hydrants during cold weather. This may be accomplished by "sounding" or striking the hand over one of the open outlets. Water or ice in the barrel shortens the length of the "organ tube" and raises the pitch of the sound. With experience, the presence of ice or water can be detected. Water must be pumped out and defective drains or valves repaired. Using salt or antifreeze in the barrels has limited value in preventing freezing and the corrosive effect may impair the operation of the hydrant. If the hydrant is only slightly bound by ice, tapping the arm of a wrench on the nut may release the stem. Moderate blows should be used to prevent breaking the valve rod. Water or ice in the barrel may sometimes be detected by lowering a weight on a stout cord into the barrel.

(f) Check hose houses, if provided at hydrants, to see that hose and equipment are in place and not tampered with.

WATER SOURCES

Sources of water supply include surface lakes, rivers, and impounded supplies. Their supply often is unlimited, but can be hampered by ice in freezing weather, drought, and evaporation, if there is no supplementary supply.

Ground water supply includes wells and springs, each of which may have limitations. Surface and ground supplies may be used in drafting operations for pumpers, or for permanent supply to an underground main and one or more hydrants.

Gravity flow and direct pumping are used for water distribution, and sometimes these also supply storage, by dumping excess water automatically into the storage facility.

Large municipal water systems sometimes are supplied by one or more reservoirs, operated by the state or county, or may have their own surface and ground sources. The quantity of domestic water required is determined by studying rates of consumption over several years and determining the average daily, maximum daily, and peak hourly consumption. To this must be added the quantity of water needed for public fire protection.

Pressures in the range of 65 to 75 psi are best in most municipal systems, and a minimum residual pressure of 20 psi should be maintained at hydrants that are delivering the required fire flow. Piping and fittings in public water systems are designed for maximum working pressures of 150 psi, but it is not good practice to operate at that level.

A number of formulas have been used for determining fire flow in municipalities, but there are many variables that influence the calculations. Some appropriate publications are listed at the end of this chapter.

Private water systems often are connected to public systems to provide water to sprinkler and standpipe systems, open sprinklers, yard hydrants, fire pumps, and private storage reservoirs or tanks. It is important that this supply be separate from domestic and industrial service within the property. Separation is accomplished by inserting a check valve between the public and private distribution systems. This valve will permit the flow of public water into the private system, but will prevent flow in the opposite direction. Often, the water supplied to a private system by the municipality is metered. Some meters are equipped with a bypass feature that provides an unobstructed waterway when the demand is heavy, such as when fire suppression systems are operating. Dry barrel hydrants are used in most areas susceptible to freezing and the wet barrel type is sometimes used where the temperature remains above freezing.

Yard hydrants have gate valves and indicating valves, the latter including an underground gate valve with indicator post; an underground butterfly valve with indicator; and an outside screw and yoke (O.S.& Y.) gate valve in a pit. The posts carry a metal plate showing what they control. The proper direction for turning should be indicated. Posts are often locked in the open position, because they may be closed

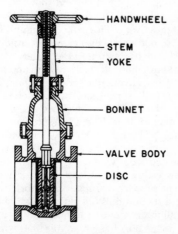

— HANDWHEEL

— STEM
— YOKE

— BONNET

— VALVE BODY

— DISC

Figure 27-1. A typical outside screw and yoke (O.S. & Y.) valve.

by mistake or vandalism at a time when the water is critically needed for a fire.

STORAGE FACILITIES

Three types of containers are used for water storage in plant properties: gravity tanks, ground tanks and pressure tanks. Each serves a different function.

Gravity tanks are made of steel or wood and range in capacity from 5,000 to 500,000 gallons (steel) and 5,000 to 100,000 gallons (wood). Depending upon requirements, they are raised on an independent steel tower so that the tank's bottom capacity line is from 75 to 150 feet above ground. They are designed to resist earthquakes and other pressures; so are suction tanks. Reinforced concrete towers are also used and tanks have been placed upon the buildings or other structures they supply. Towers are designed to support the dead load (weight of the structure and fittings) and the live load (weight of the water when the tank is filled to overflowing), plus allowance for weight of ice and snow.

Provision must be made for heating the tank so that ice will not block water delivery. This can be accomplished by gravity circulation of hot water, by steam coils inside the tank, or by direct discharge of steam into the water.

Suction tanks are made of steel or wood and are set upon a ground foundation of concrete, crushed stone, or sand. The foundation is usually surrounded by a concrete ring wall. A pit measuring 6 by 9 feet and 7 feet deep houses valves, tank heaters and other fittings. There is appropriate waterproofing and drainage. There should be a water level indicator or a high and low water electrical alarm. The gage is normally installed in a heated room where it is readily accessible. There should be an overflow pipe at least 3 inches in diameter.

Embankment supported, rubberized fabric tanks are also used as ground suction tanks. They are available in 20,000- and 50,000-gallon sizes, and in 100,000-gallon increments up to 1 million. Such a tank has a reservoir liner with an integral flexible roof and is designed to be supported by earth on the bottom and on four sides. Water temperature must be maintained at not less than 42°F. This can be done with a heat exchanger and water circulation system.

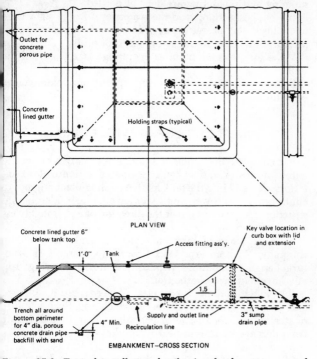

PLAN VIEW

EMBANKMENT—CROSS SECTION

Figure 27-2. Typical installation details of embankment-supported rubberized-fabric tanks, including fittings.

Pressure tanks are used to supply sprinkler and standpipe systems, hose lines, and water spray systems. They are sometimes connected to fire pumps and gravity tanks. Tank capacity ranges from 2,000 to 9,000 gallons — at least 4,500 gallons for ordinary hazard occupancies. The tank normally is kept about two-thirds full of water, with an air pressure of

75 psi. Tanks should be housed in noncombustible structures unless they are in a heated room within a building. Tank interiors are inspected every three years for corrosion, and scraping and repainting are done as needed.

In your inspection, read the water level in the sight gage and check the pressure gage. If there is a recording of pressures examine the range for excess of limits.

Make sure that the pump is operating correctly to fill the tank.

Examine the air compressor for capacity and maintenance condition.

INSPECTION OF GRAVITY TANKS

For complete inspection of a tank it is usually necessary to climb the tower and, when checking condition of tank interiors, descend into the tank itself. You should not attempt to climb tanks and towers until you have had considerable instruction and practice under the direction of a thoroughly experienced person.

(a) Note name of tank manufacturer and the installing company. With an experienced company, features of structural design, foundations, wind and earthquake loadings usually have had proper attention.

(b) Note if tank site is kept clear of weeds, brush, rubbish, and piles of combustible material, which might cause failure of steelwork through fire or corrosion. Note if tank is safe from damage by fires in nearby buildings.

(c) Note any tank use other than for fire service.

(d) Read mercury gage or other water level indicator and consult records that are kept of such readings throughout the year.

(e) Tank should have an overflow. Ask that tank be overflowed to test water level indicators and tank filling arrangements.

(f) Note general maintenance of the tank structure, tank, and accessories.

(g) Ask how recently the owner has had the tank completely gone over by an experienced tank contractor. Where water conditions make frequent cleaning necessary, ask the last date when tank was cleaned.

(h) Consult records kept of weekly valve inspections. Note if valves were found wide open and properly sealed. Have each valve given operating test.

(i) Inspect valve pit. Construction and arrangement should be satisfactory with adequate clearance around pipes, and valves in the pit, manhole, and ladder should be in good repair. The pit should be waterproofed and drain properly.

(j) Consult daily records, which should be kept, of thermometer readings of tank water temperature during cold weather. Ask if heating system is checked daily in cold weather and when plant is closed.

(k) Ask about any experiences with ice on any part of the tank structure.

(l) Inspect the tank heating system, particularly any separate above-grade heater house. Note construction of heater house. Determine that roof will properly support frostproof casing and any other loads imposed.

(m) Review the records of tank painting, and estimate condition of painting. Note if paint surface inside tank has been checked within two years.

(n) If cathodic protection is provided in a steel tank, determine when the equipment was given its last maintenance inspection by the supplier.

FIRE PUMPS

The centrifugal pump is used almost universally to provide water under pressure for industrial plant fire protection. Horizontal and vertical fire pumps are available in capacities up to 5,000 gpm, with pressures ranging from 40 to 200 psi for horizontal pumps and 75 to 280 psi for vertical pumps. These pumps may be single- or multi-stage, depending upon the number and arrangement of the impellers. Pump capacity is rated on the discharge of one stage in gallons per minute; pressure rating is the sum of the pressures of individual stages, minus a small head loss.

Pump performance is recorded in a series of head-discharge curves under test conditions. There are three limiting points: the shutoff, with the pump operating at rated speed with the discharge valve closed — the total head at shutoff should not exceed 120 percent of the rated head at 100 percent capacity; rating — the curve should pass through or above the rated capacity and head; and overload — at 150 percent of rated capacity, the total head should not be less than 65 percent of rated total head.

Horizontal shaft centrifugal fire pumps should be operated under positive suction head, especially with automatic or remote control starting. If the location requires suction lift, try a vertical turbine-type pump.

The vertical pump has the capacity to operate without priming, and it is often used in streams, ponds and pits. Such a pump would consist of a motor of right-angle gear drive, a column pipe and discharge fitting, a drive shaft, a bowl assembly housing the impellers, and a suction strainer. Its operation is similar to a multi-stage centrifugal pump. Vertical pumps have the same standard capacity ratings as horizontal pumps.

When inspecting any fire pump, first read the manufacturer's rating data and note any exceptions to normal test and operating procedures. Have the pump started and watch for signs of leakage, overheating and irregular performance.

Note if the pump is aligned correctly with the driving motor or turbine, and if there is leakage at the stuffing box glands.

Watch pressure gages for erratic performance, which may indicate poor suction, obstructions, inadequate water supply, or insufficient immersion of the suction pipe.

Close all outlets, including the relief valve, and note if the pump shuts off at the correct pressure.

If the pump is supposed to start automatically, try this by opening a test connection.

Fire pumps are supposed to be tested annually, so ask for the records of the most recent tests, which should record performance of the pump, driving engine, suction, and power supply and show the results of flow tests.

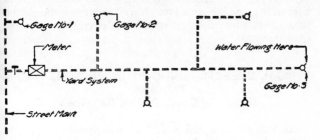

Figure 27-3. Gage connections for yard system flow tests.

METHOD OF MAKING FLOW TESTS ON YARD SYSTEM

The diagram shows a yard system fed by a city connection, which may be used as an example of how water flow in such a system is checked. Flow is measured as described in Flow Test of Water Main to Determine Water Available for Sprinklers, Chapter 28. Friction losses in pipes may be estimated from the table, Friction Loss in Pipes, in this chapter.

Gage No. 1 shows street pressure.

Gage No. 2 is located on a branch line as near to the street connection and meter as possible. The amount by which the reading of gage No. 2 is different from that of gage No. 1 shows the loss in the connection. Meters frequently cause large pressure drops.

Gage No. 3 gives readings at the point where water is flowing during a test. The difference in pressure should be explained by the expected friction losses in the piping system between gage No. 3 and gage No. 2.

The pressure for a specific flow at the end of the system at gage No. 3 should be the same as for the same flow at gage No. 1 less calculated losses in the piping between No. 1 and No. 3. If observed pressure is less than calculated, take readings at other outlets between No. 1 and No. 3 and make calculations of pressure to be expected for a given flow until the obstruction is located. With tests such as described, a hydraulic grade line for a given run of piping may be plotted which will often show otherwise undetectable sources of pressure drop.

TYPICAL DEFECTS FOUND BY INSPECTORS' FLOW TESTS OF UNDERGROUND PIPING SYSTEMS

The value of a careful check by the inspector of a private system of underground piping is shown by the following list of some defects that have been found in systems.

(a) Connection never made to city main.

(b) Connection to city main made, but gate valve never opened or only partly open.

(c) Broken meters or clogged meter fish traps.

(d) Mains smaller than indicated on plans.

(e) Mains with seriously reduced areas because of sedimentation or hard deposits on the inside.

(f) Mains entirely closed with mud.

(g) Serious leakage of underground systems.

(h) Frozen mains or valves.

(i) Inoperative hydrants.

(j) Valves entirely or partly closed due to ignorance of employees as to which way to turn the valve or due to careless operation.

(k) Check valves pointing in the wrong direction.

(l) Leaky check valves.

(m) Existence of valves and meters not known.

(n) Blocked suction on fire pump.

(o) Frozen drop pipe on gravity tank and other frozen tank connections.

SI UNITS

The following conversion factors are given as a convenience in converting to SI units the English units used in this chapter.

$$1 \text{ in.} = 25.4 \text{ mm}$$
$$1 \text{ ft} = 0.305 \text{ m}$$
$$1 \text{ psi} = 6.895 \text{ kPa}$$
$$1 \text{ gal} = 3.785 \text{ litres}$$

BIBLIOGRAPHY

McKinnon, G.P. ed., *Fire Protection Handbook*, 15th ed., National Fire Protection Association, Quincy, MA, 1981. Chapters 1 through 8 of Section 16 cover aspects of industrial fire protection including water and water additives, hydraulics, water supply requirements, water distribution systems, water storage and suction supplies, fire pumps, fire streams, and testing water supplies.

NFPA Codes, Standards, and Recommended Practices. (See the latest *NFPA Codes and Standards Catalog* for the availability of current editions of the following documents.)

NFPA 13, *Standard for the Installation of Sprinkler Systems*. Explains the requirements for installation, water supply, and performance of sprinkler systems.

NFPA 13E, *Recommendations for Fire Department Operations in Properties Protected by Sprinkler and Standpipe Systems*. An explanation of fire control techniques.

NFPA 15, *Standard for Water Spray Fixed Systems for Fire Protection*. Covers systems components, required water supply, design and installation, tests, maintenance, automatic fire detection, and other essentials.

NFPA 24, *Standard for the Installation of Private Fire Service Mains and Their Appurtenances*. Explains requirements for water supplies, valves, hydrants, hose houses and equipment, master streams, underground pipes and fittings, and rules for laying pipes.

NFPA 26, *Recommended Practices for the Supervision of Valves Controlling Water Supplies for Fire Protection*. Explains methods of supervision, testing, emergency sealing, and precautions when equipment is shut off for repairs.

Chapter 28

AQUEOUS EXTINGUISHING SYSTEMS

Because water is an efficient and universally available extinguishing agent, the use of water in automatic extinguishing systems is continually increasing. Sprinkler systems for protecting dwellings are now becoming practical, and newer and more complex systems are being designed for industrial plants and other large occupancies. In fact, the hazards of a particular building are an important factor in determining what kind of fire protection is needed for that building.

For simplicity, occupancies are classified into three types — light hazard, ordinary hazard, and extra hazard.

The *light hazard class* includes buildings like dwellings, apartments, churches, hotels, schools, offices and public buildings, where the quantity of combustible materials is low and fires are expected to have a low rate of heat release.

The *ordinary hazard class* includes ordinary mercantile, manufacturing, and industrial properties, and these are considered in three groups — one in which combustibility and quantity of materials is moderate, stockpiles do not exceed 8 feet in height, and fires are expected to release heat at a moderate rate; the second group includes properties where the quantity of combustible products is moderate, stockpiles do not exceed 12 feet in height, and fires with moderate rates of heat release are expected. In this group might be cereal mills, textile and printing plants, and shoe factories. The third group includes occupancies with high fire potential, such as flour mills, piers and wharves, paper manufacturing plants, storage warehouses and others.

In the *extra hazard class* are two main groups. One includes occupancies that contain little or no flammable or combustible liquid, but may have severe fires. These would include die casting, metal extruding, sawmills, rubber production, and upholstering operations using plastic foams. The other group would include asphalt saturating, flammable liquid spraying, open oil quenching, solvent cleaning, varnish and paint dipping.

In addition to these hazard groupings, there are *special occupancy conditions* which require consideration, such as high-piled combustibles, a variety of flammable and combustible liquids, combustible dusts, chemicals, and explosives. For some of these, water may not be effective unless combined with an additive, but for most occupancies water can be the primary extinguishant. Aqueous extinguishing systems can be designed to meet a variety of situations, but they must be supplied and maintained in good operating condition.

Brief descriptions of the principal kinds of sprinklers are included in the preceding chapter. This chapter deals with inspection and maintenance of these systems, and related information.

There are two occasions when inspectors must be most observant and critical — when a new sprinkler system is being installed, and when the installed system must be shut down for repairs or extension. On each of these occasions, the occupancy will be lacking its principal security of automatic fire detection and extinguishment. Many serious fires have resulted in these situations, sometimes from carelessness or by accident. It is highly important that the work areas for installation be kept clear of unnecessary combustibles, that portable fire extinguishers be readily available, and that the plant fire brigade or local fire department be notified of the times of shutdown.

As an inspector you should determine whether the building which you are inspecting is of light-, ordinary-, or extra-hazard classification, and accordingly, determine the fire protection requirements. This will help you judge the extent and capability of the automatic extinguishing system. Most of the flaws or errors will be fundamental and easy to find; others may be discovered in routine tests.

WET-PIPE SYSTEMS

Where wet-pipe sprinklers have not worked well in fires

there were two reasons — the sprinkler water main was blocked or otherwise impeded, or someone closed the valve on the supply line. Therefore, one of your first acts of inspection is to verify that sufficient water is available to the sprinkler system. The sketches, maps, and other data developed before and after the system was installed should indicate the flows and pressures throughout the system, the piping layout, and the location of valves. You should arrange to have each valve operated under your observance to verify that it performs well. Also note if there is a sign or other indication of whether the valve should remain open or closed.

The outside fire department connection merits your next attention. Remove the outlet caps, examine the threads, and then the interior for rags or other blockage. Note if the threads are compatible with those of the local fire department or if adaptors are needed.

If there are yard hydrants and indicator post valves in the outside area, check these for condition and ease of operation. Determine if O.S.&Y. valves are supposed to be locked in the open position. Make sure the valve pit is kept clean and well maintained.

If there are underground valves, make sure that the appropriate keys or wrenches are readily accessible. If municipal hydrants are on the property, check with the local fire department on how they will be operated during a fire.

There are central station and proprietary supervisory services which provide continual surveillance of valves in water systems. You should determine if the building or plant you are inspecting has this service and, if so, the extent and effectiveness of the valve suppression.

Reports on valve conditions should be made weekly or at some other regular interval and should identify: the valve number; whether it is open or closed; whether it is sealed properly, or locked; whether it is in good operating condition, and does not leak; its accessibility; and if the valve key or wrench is in place.

Sprinklers should be examined next. They should be clean, not covered with dust, grease, or paint, particularly the heat-responsive element. They should not be obstructed by light fixtures, stored materials, or the movement of doors or windows. Corrosion is another impediment to sprinkler operation.

All gage readings of the system should be recorded and compared with the sketch information and other data.

There are means for testing alarm valves and water flow devices and you should arrange for such tests. Drain valves and test pipes should also be operated during an inspection.

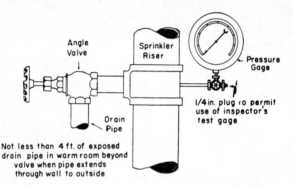

Figure 28-1. Tests and drain connection for a wet-pipe sprinkler system riser.

DRY-PIPE SYSTEMS

Dry-pipe systems differ from wet-pipe systems in several ways. First, the water is held from reaching the sprinklers by the pressure of compressed air or nitrogen. The water is released only after the dry-pipe valve is actuated; then it flows instantly to all sprinklers.

Another difference is in the use of pendant sprinklers; in an area subject to freezing the dry pendant type may be used, but standard pendant sprinklers are allowed on return bends in heated areas.

Check valves may be installed in branches of a dry-pipe

system and indicating drain valves are connected by a bypass around each check valve to allow draining of the system.

Dry-pipe systems with capacity of more than 500 gallons must have quick-opening devices, which are placed close to the dry-pipe valve. There will also be a soft disc globe or angle valve in the connection between the dry-pipe sprinkler riser and the quick-opening device.

The dry-pipe valve and the supply pipe must be protected against cold and mechanical injury. A low differential dry-pipe valve, installed above the clapper to prevent water accumulation, can also serve as a high-level signaling device or an automatic drain.

Air pressure from the system may come from a shop system or an automatic compressor. The pressure should be that listed on the instruction sheet for the dry-pipe valve, or 20 psi above that valve's calculated trip pressure. The same requirement applies to nitrogen.

Pressure gages are installed on the water and air sides of the dry-pipe valve, at the air pump, at the air receiver, in each independent pipe from the air supply to the system, and at exhausters and accelerators. You can identify these gages and their purpose from the system installer's plans or the manufacturer's literature.

Your inspection procedure will be much like that for a wet-pipe system, except that you will be more concerned with the air pressure and the devices it actuates.

PREACTION AND DELUGE SYSTEMS

These systems may be used individually or in combination, with or without automatic supervision for the sprinklers or the fire detection devices. Your inspection will be concerned wtih the condition and functioning of each system.

As mentioned in Chapter 27, each of these systems uses air to retain water until the system is actuated, but in a preaction system, only one or a few sprinklers may open, while in a deluge system all sprinklers discharge.

Things to look for:

At least two spare fusible elements of heat-responsive devices for each temperature rating must be available for replacement purposes.

The release system might be actuated hydraulically if the fixed temperature or rate-of-rise heat defectors are located at extreme heights above the deluge valve or valve actuator.

Check the location and spacing of fire detection devices and the functions of the system. Refer to fire protection standards and the manufacturer's literature.

If there are more than twenty sprinklers on a preaction or deluge system, the sprinkler piping and fire detection devices must be supervised automatically.

A system with more than 1,000 closed sprinklers must have more than one preaction valve.

If fire detection devices in circuits are not accessible, additional devices for testing must be in an accessible location for each circuit, and shall be connected to the circuit at a point that will assure an adequate test.

The system should include testing apparatus that will produce sufficient heat or impulse to operate any normal fire detection device. In hazardous locations, where explosive materials or vapors are present, hot water, steam, or other nonignition method may be used.

Combined dry-pipe and preaction systems must be constructed so that either system will still operate if the fire detection system fails, and the fire detection system will operate if either of the sprinkler systems fail.

WATER SPRAY PROTECTION

A system of water spray protection is very similar to a sprinkler system, except for the pattern of spray discharge. Water spray is used for extinguishment, controlled burning, exposure protection, and preventing fire. It is used for ordinary combustibles, for electrical equipment, flammable gases and liquids, processing equipment and electrical cable trays and runs.

These systems are designed to provide optimum control, ex-

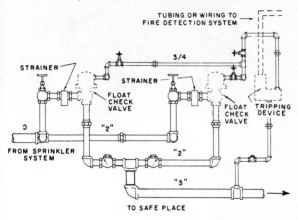

Figure 28-2. Arrangement of air exhaust valves for combined dry-pipe and preaction sprinkler system.

tinguishment, and exposure protection for special situations; they are not a replacement for sprinkler systems. They can be designed for a variety of discharge volumes and patterns, including ultra-high-speed response in milliseconds. The type of system will depend on the extent of hazard and required water discharge.

These systems commonly are used to protect tanks of flammable liquids and gases, electrical transformers, oil switches, rotating electrical machinery, wall openings, and similar fire problems. They consist of fixed piping supplied by sufficient water, and water spray nozzles of specific design for the discharge and distribution over the area to be protected. Water flow is started manually or automatically, usually by actuation of detection equipment. Water spray systems have heavy demand for water because high density discharge from many nozzles is often needed. The maximum capacity for a single system is recommended to be 3,000 gpm.

Automatic fire pumps and adequate water supply from underground mains or elevated gravity tanks are other requirements.

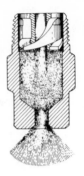

Figure 28-3. Water spray nozzles having internal spiral passages. (Spraying Systems Co., Type H, and "Automatic" Sprinkler Corp. of America, Type MA)

These systems must be inspected and maintained on a regular basis. Items to be inspected include strainers, piping, control valves, heat actuated devices, detectors, and spray nozzles, especially those having strainers. It is important that small water passages be kept clear. If spray nozzles are subject to paint vapors and other coatings, blowoff caps can be used to protect against foreign matter and corrosion.

FOAM EXTINGUISHING SYSTEMS

Fire fighting foam is a combination of water and concentrated liquid foaming agent. It floats on the surface of flammable and combustible liquids, and makes a covering that excludes air, cools the liquid, and seals the layer of vapor. It also forms a resistant blanket over transformers and other irregular shapes to smother flames. There are several kinds of foaming agents, and their effectiveness varies with application and the properties of the fire-involved fluid or material.

Foam may be applied by portable devices or fixed extinguishing systems. In either application, the compound must be of the right mixture and the application must be con-

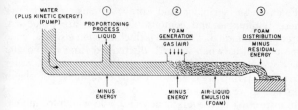

Figure 28-4. The steps in air foam generation.

tinuous and consistent. Foam breaks down and vaporizes its water content in direct exposure to heat and flame, but if applied in sufficient volume, it can overcome this loss and diminish or extinguish the fire. The smothering layer can be broken and dispersed by mechanical or chemical action, or turbulence from air or fire gases. Nevertheless, it can be applied with efficiency by automatic extinguishing systems.

For the most effective application, the following conditions must exist. The fire-involved liquid must be below its boiling point at ambient conditions of temperature. Foam must be applied carefully at temperatures above 212°F, when it will form a mixture of steam, air and fuel, and may increase four-fold in volume. The burning liquid must not be unusually destructive to the foam, and the foam must not be highly soluble in the liquid. The liquid must not be reactive to water, and the fire, preferably, must be on a horizontal surface. Foam can diminsh fires on vertical surfaces, but it must form a consistent blanket for full effect.

Types of foaming agents include: aqueous film-forming, fluoroprotein foaming, protein foaming, high expansion foaming, synthetic hydrocarbon surfactant foaming, low temperature foaming, "alcohol-type," chemical agents and powders. The last are practically obsolete.

Foam can be applied by manual hose streams or by fixed equipment. It can be effective on the surface of burning fuel or beneath the surface. In fixed equipment, it can be applied as ordinary or high expansion foam, or in subsurface injection.

Foam-water sprinkler and spray systems are effective where flammable and combustible liquids are processed, stored, and handled. The foam is discharged in essentially the same pattern as when water is discharged from the nozzle. These systems are used to protect aircraft hangars, oil-water separators, pump areas and oil piping manifolds, petroleum piers and other installations.

In inspecting such equipment you should watch for indications of corrosion, clogging of orifices, sticking of valves, and malfunction of electrical parts. You should also review test results.

The foam system should meet the requirements of fire protection standards and otherwise show evidence of regular inspections and maintenance. Regular tests should confirm the following: dimensions and patterns of the discharge pattern; percent of foam concentrate in the finished solution; degree of foam expansion in the finished compound; rate at which water drains from the foam after discharge; and film-forming capability of the foam concentrate.

The mechanical equipment of the system should be inspected like a sprinkler system.

STANDPIPES AND HOSE SYSTEMS

Standpipe and hose systems provide a means for the manual application of water to fires in buildings. They do not, however, take the place of automatic extinguishing systems. They are always needed where automatic protection is not provided and in areas of buildings not readily accessible to hose lines from outside hydrants.

Water Supplies

The water supply needed depends on the size and number of streams to be used and the length of time they may have to be operated, as well as consideration for automatic sprinklers using the same riser. Acceptable supplies include municipal systems in which the pressure is adequate, automatic fire pumps, manually controlled fire pumps with pressure tanks, pressure tanks, gravity tanks, and manually controlled fire pumps operated by remote control devices at each hose station.

Types of Systems

There are four types of standpipe systems. The most desirable type of system is one that is charged at all times. Another is a dry-pipe system equipped with remote control devices at each hose station that admit water into the system. Third is the dry-pipe system for unheated buildings. In this system, a dry-pipe system for unheated buildings. In this system, a dry-pipe valve prevents water from entering the system until the stored air pressure in the discharge side falls below the water supply pressure. Finally, there is the dry standpipe and hose system that has no permanent water supply. This type receives its water from a fire department pumper.

Inspections

Periodic inspection of all portions of standpipe systems is essential. See that tanks are filled, and where pressure tanks are employed, that the pressure is at least 75 psi. Check that the valves in automatic sources of water are open. Also test the electromechanical supervision of such valves. Examine the threads at the fire department connection, and be sure that the waterway is not clogged with foreign material.

At each discharge outlet or hose station, check the valve for leakage, and examine the hose threads. Where hose is provided for occupant use, check the condition of the hose and nozzle and see that the hose is properly stored.

STEAM SMOTHERING SYSTEMS

These systems were used many years ago but are rare today

and are not recommended for fire protection. They may be used to protect ovens or cargo spaces on ships. An accepted application is 8 pounds of steam per minute for each 100 cubic feet of oven volume.

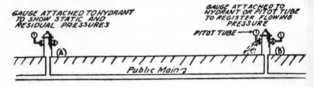

Figure 28-5. Test setup for flow testing water main.

FLOW TEST OF WATER MAIN

The following test may be made, using two hydrants nearest to where a proposed connection from a main is to be taken for sprinklers. It is accurate enough to show that there is the usual minimum of 500 gpm of water available with a flowing (or residual) pressure sufficient to provide 15 psi or more at the top line of sprinklers. Permission for the test should be secured from the water utility in the case of public mains.

1. Attach gage to hydrant (A) and obtain static pressure.
2. Attach second gage to hydrant (B) and remove cap from one of its 2½-in. outlets.
3. While hydrant (B) is uncapped, measure the diameter of the outlet to check its size. While usually the inside diameter of the hydrant opening is close to 2½ in., some butts are sufficiently different to require a fairly close measurement which for convenience should be taken to the nearest hundredth of an inch.
4. Open hydrant (B) and read the pressure on both hydrants.
5. Note if the full opening is filled with the water flowing. Full flow is needed for accurate measurement.

Accuracy of the test is slightly better if, instead of a gage on hydrant (B), the pressure is read on a gage on a pitot tube held in the flowing stream. The position of the tube in the stream should be varied to get the most representative reading. In most openings the pitot gage reading will be best with the tube near the middle of the stream.

If accurate results are needed, attach a 50-foot length of $2\frac{1}{2}$-inch cotton rubber-lined hose to each of the $2\frac{1}{2}$-inch connections on hydrant (B). Each hose line should have a playpipe and a smooth cone nozzle tip $1\frac{1}{8}$ inch or larger. Both lines are turned on together, and pressure at the nozzles is read with a pitot tube and gage. Discharges can be measured quite closely with good nozzles, and the total discharge is the sum of the discharge computed for each nozzle. Measure the nozzle diameters with a scale that can be read to hundredths of an inch. If it is desired to know how much water is available at a selected residual pressure, take readings at two or more flows and plot a flow curve.

Discharge is computed from the formula or estimated from the *Table of Theoretical Discharge Through Circular Orifices*.

The static pressure difference at the top line of sprinklers may be computed by deducting from the minimum reading of the gage on hydrant (A), 0.434 pound for each foot the highest sprinklers are above gage.

THEORETICAL DISCHARGE THROUGH CIRCULAR ORIFICES

From Fire Stream Tables, originally prepared by John R. Freeman.

Discharge from hydrant opening or nozzle is computed from the formula:

$$Q = 29.83 \, cd^2 \sqrt{p}$$

where Q is gallons per minute, c is a coefficient of discharge for the opening, d is its diameter in inches and p is the pressure read at the gage on the hydrant flowing (or pitot tube gage) in pounds per square inch.

Pressure, psi	Head, ft	1	1⅛	2½	4
			Flow, gpm		
15	34.6	116	146	722	1,849
30	69.3	164	207	1,022	2,616
45	104	200	253	1,253	3,203
60	138	231	293	1,445	3,700
90	208	283	358	1,770	4,531
120	277	327	414	2,044	5,232

The values in the above table are computed from the above formula with $c = 1.00$.

PRESSURE TEST ON SPRINKLER SYSTEM

1. Note pressure indicated by gage (usually located just above the indicator gate valve controlling the system). As a rough rule, water pressure readings should be at least 30 psi for one-story buildings plus 5 psi for each additional story.

2. Open main drain valve or other outlet and allow full flow of water for at least 2 minutes. Tap the gage lightly to be sure pointer is moving freely, and note the pressure again. This is the flowing pressure. The pressure at the top line of sprinklers would be less. The static pressure difference may be calculated (deduct 0.434 psi for each foot the top sprinklers are above the gage).

3. Close drain valve again and note the pressure indicated on gage. The opening of an outlet usually causes a reduction in the pressure reading on the gage. Noting the amount of this drop and the subsequent recovery of pressure reading upon closing of the outlet serves as a partial test as to the water supply being fully turned on; it may also indicate any unusual obstruction in the piping. The test is valuable for comparison with tests made on the public water mains, and flow test made on the yard system.

4. Attach an inspector's gage and note pressures in com-

parison with those indicated by the permanent gage. (This is a test of accuracy of gages.)

5. Make record of the test with date for reference at future inspections.

INSPECTION OF HOSE THREADS

An inspection of threaded couplings for hose connections should first look for evidence of injury or damage. For example, see whether coupling makes up by hand and that the swivel operates freely.

The waterways should be measured. An inspector's pocket scale, graduated in inches and tenths, is suitable for this. Measurement with the scale in three or four different positions will determine the full diameter and that the fitting is not out of round.

Inspectors who may have frequent occasion to measure hose threads and couplings will find it useful to have a pocket slide caliper for inside and outside measurements and screw pitch gages at least for standard 2½- and 1½-inch threads.

Better equipment for inspection of threads are plug and ring gages. More accurate than a coupling for inspectors' field work are straight "go" ring gages for 2½-inch and 1½-inch standard threads and a straight "go" plug gage for 2½-inch. The former are for identifying threads on male outlets as on standpipes; the latter is for female couplings on hose and siamese connections.

INSPECTION OF WOVEN-JACKET RUBBER-LINED FIRE HOSE

(a) Determine that the proper amount of hose is located where needed, in hose houses or on hose reels, racks, or carts.

(b) Determine that hose is not stored where it is subject to damage from mechanical injury, heat, mildew or mold, acid, gasoline or oil, and that it does not show evidence of having been exposed to these other than in the normal wear and tear of use.

(c) Is hose adequate quality? Evidence that each length passed suitable tests when new is given by listing marks of testing laboratories.

(d) After use at fires or drills, is hose brushed and washed and scrubbed with plain water? Soap may be used where hose has been exposed to oil but should be carefully rinsed off.

(e) Hose not used should be taken out periodically and have water run through it periodically. When was this last done? It should be done two or three times a year. After cleaning and drying it should be rolled and stored, or if in a hose house or hose wagon, it should be repacked with the folds in a different place than before.

(f) Are there suitable facilities for drying hose, so that it does not have to be laid out in the sun on concrete driveways or sidewalks? Suitable facilities are hose drying cabinets, hose racks or hose towers.

(g) Has the hose been given a pressure test within a year?

(h) Are couplings standard thread, and will they connect with other hose and hose connections and hose gates, with the public hydrants and fire apparatus, and with the hose equipment of nearest neighbors likely to give aid? If connections have to be made at any time with nonstandard threads, have a number of suitable adapter couplings been provided?

(i) Examine couplings to see that they are not out of round or have injured threads. See that they may be coupled by hand (couplings should not be oiled, as oil damages the hose). Dirty couplings may be spun in a pail of soapy water. Are coupling gaskets intact and proper size (so they do not project into the flowing stream)?

TESTING WOVEN-JACKET RUBBER-LINED HOSE

Regular service tests should be made annually of all hose. Hose at hose houses, or where it will ordinarily be used in short lines, may be tested to 150 psi, and double-jacket hose on fire brigade trucks and fire department apparatus should be tested to 250 psi. "Proof" pressure is 50 to 100 percent above the usual annual test or working pressures and is a pressure to which new hose is subjected. The pressure that hose will withstand without bursting is much higher. Burst pressures are as follows corresponding to respective "proof" pressures as shown in parentheses:

Trade Size, Inches, and Number of Jackets	Burst (and "Proof") Pressures Pounds per Square Inch
1½, 2, 2½ Single	500 (300), 600 (400), 750 (500)
1½, 2, 2½, 3, 3½ Multiple	600 (400), 900 (600)

Tests are made of hose with couplings to assure that couplings are properly attached. When other hose test facilities are not available, a convenient method of test is to use a fire department pumper and hook up each length of hose with a shut-off nozzle on the end. The line is laid out straight, not pulled, without twist. With the nozzle open, water is turned on and allowed to flow until flowing freely and all air has cleared. The nozzle is closed slowly and the test pressure is built up by the pump. The pressure gage on the pumper is observed closely, so that if a length bursts the pressure at which failure occurred may be noted.

The test pressure should be maintained for 3 minutes. It is important that the nozzle be open when water is turned on as the sudden release of entrapped air, as by failure of the hose or blowing off of a coupling, would be violent and could cause injury. It is a good idea to lash the nozzle during the hose testing.

SI Units

The following conversion factors are given as a convenience in converting to SI units the English units used in this chapter.

$$1 \text{ ft} = 0.305 \text{ m}$$
$$1 \text{ lb} = 0.454 \text{ kg}$$
$$1 \text{ psi} = 6.895 \text{ kPa}$$
$$\tfrac{5}{9} (°F - 32) = °C$$
$$1 \text{ gal} = 3.785 \text{ litres}$$
$$1 \text{ cu ft} = 0.0283 \text{ m}^3$$
$$1 \text{ gpm} = 3.785 \text{ litres/min}$$

BIBLIOGRAPHY

McKinnon, G.P. ed., *Fire Protection Handbook*, 15th ed., National Fire Protection Association, Quincy, MA, 1981. Section 16 deals with water supplies for fire protection. Section 17 covers water-based extinguishing systems, and Chapter 4 of Section 18 discusses foam extinguishing agents and systems.

NFPA Codes, Standards, and Recommended Practices. (See the latest *NFPA Codes and Standards Catalog* for the availability of current editions of the following documents.)

NFPA 13, *Standard for the Installation of Sprinkler Systems*. Explains the full requirements for various types of sprinkler systems, including inside and outside protection, water supplies, system components, types of systems, and specialized units like hydraulically designed and high-rise systems.

NFPA 14, *Standard for the Installation of Standpipe and Hose Systems*. Describes the required components and delivery systems.

NFPA 15, *Standard for Water Spray Fixed Systems for Fire Protection*. Covers system components, design, installation, tests, hydraulic calculations, and automatic detection equipment.

NFPA 16, *Standard for the Installation of Deluge Foam-Water Sprinkler Systems and Foam-Water Spray Systems*. Covers regular and spray systems, from supply through testing and maintenance.

NFPA 22, *Standard for Water Tanks for Private Fire Protection*. Contains detailed information on elevated and ground level tanks and water supply.

NFPA 24, *Standard for the Installation of Private Fire Service Mains and Their Appurtenances*. Explains the needs and requirements for sprinklers and other outside measures.

Chapter 29

NONAQUEOUS EXTINGUISHING SYSTEMS

Nonaqueous extinguishing systems are those fire suppression systems that do not rely upon water as a major element in the extinguishment process. Namely, they include carbon dioxide systems, dry chemical systems, and halogenated extinguishment systems.

CARBON DIOXIDE SYSTEMS

Carbon dioxide has a number of properties that make it a desirable fire extinguishing agent, especially in situations where water is not the answer. It is noncombusible, it does not react with most substances, and it provides its own discharge pressure. As a gas, carbon dioxide can penetrate the fire area, and it leaves no residue, thus facilitating clean-up.

Extinguishing Mechanism

Carbon dioxide is effective as an extinguishing agent primarily because it smothers the fire by diluting oxygen in the vicinity of discharge to the point where it will no longer support combustion. In other words, carbon dioxide, which does not displace combustion, displaces oxygen, which does support combustion. In addition carbon dioxide has a cooling effect which is helpful, especially when it is applied directly to burning material.

Limitations

Carbon dioxide does have its limitations as an extinguishing agent. Its cooling abilities are not used to best advantage because the agent does not wet or penetrate the fuel and many times the areas into which the agent is discharged is inadequate for retaining an extinguishing atmosphere.

Frequently, liquid fuel fires are extinguished by discharging the agent directly onto the burning material, and a 30-second discharge is often sufficient to cool the fuel to below its reignition temperature.

Carbon dioxide is not an effective agent against fuels that contain their own oxygen supply, such as cellulose nitrate. Carbon dioxide is ineffective against reactive metals, such as sodium, potassium, magnesium, titanium and zirconium.

Hazards

The discharge of great quantities of carbon dioxide to extinguish a fire may create a health hazard. The so called "snow" that develops in the discharge may interfere with visibility, and the carbon dioxide itself may produce an atmosphere with an oxygen concentration so low that it will not support life.

Application Methods

There are a number of application methods used by carbon dioxide systems. In total flooding, the agent is applied by fixed nozzles in a manner designed to provide a uniform concentration over the entire enclosure. Local application is the discharge of carbon dioxide through a system of nozzles to a predetermined area, the object being to extinguish the fire as quickly as possible. Any adjacent area to which the fire may spread must also be covered because any residual fire could cause reignition after discharge is terminated.

Extended coverage comes into play when the enclosure is not tight enough to retain an extinguishing concentration as long as is deemed necessary.

Hand hose carbon dioxide lines are connnected to a fixed supply of carbon dioxide by means of fixed piping that can be used for total flooding or local application.

Total flooding, local application, and hand hose line

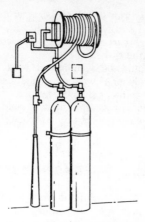

Figure 29-1. Carbon dioxide hand hose extinguishing system with hose mounted on a reel. (Walter Kidde & Company, Inc.)

systems without a permanently connected carbon dioxide supply are known as standpipe systems.

System Actuation

Total flooding and local application systems are designed to operate automatically or manually. For automatic operation, system operation is initiated by fire detection. The detection device may be any of the listed or approved devices that are actuated by heat, smoke, flame, flammable vapors or any other abnormal process conditions that could result in a fire or explosion.

Maintenance

Carbon dioxide systems must be checked frequently. Frequent scheduled visual checks should be made, and tests should be conducted at least annually to be sure that all components of the system are in proper operating condition. In order to run a satisfactory test, it may be necessary to discharge enough carbon dioxide to operate all pressure actuated parts of the system.

Cylinders should be weighed at least semiannually to detect loss by leakage. A weight loss of 10 percent of more indicates that the cylinder should be replaced. The liquid level on low pressure storage containers should be checked frequently, and the containers refilled when the loss reaches 10 percent.

HALOGENATED AGENT SYSTEMS

Halogenated extinguishing agents are hydrocarbons in which one or more hydrogen atoms has been replaced by atoms from the halogen series — fluorine, chlorine, or bromine. Halogenated agents, or halons as they are popularly called, are stored as liquids and discharged as gases and they vaporize rapidly in fire, leaving no corrosive residue. They are used to protect electrical and electronic equipment, air and ground vehicle engine compartments, and in other applications where speed or extinguishment is essential and there is a need to minimize clean-up operations.

Extinguishing Mechanism

The extinguishing mechanism of halons is not clearly understood, but there is agreement that a chemical reaction takes place which interferes with the combustion process. Bromine is more effective in combustion inhibition, or "chain breaking," than are fluorine and chlorine.

Hazards

As a result of extensive investigation of the medical effects of halons on people, maximum times of exposure to various concentrations of Halon 1301 and Halon 1211 have been established. NFPA 12A, *Halon 1301 Fire Extinguishing Systems*, permits design concentrations up to 10 percent in normally occupied areas and up to 15 percent in areas not normally occupied. Since the effective extinguishing concentration of Halon 1211 is near or above the limit for safe human exposure, Halon 1211 systems are not recognized for use in normally occupied areas by NFPA 12B, *Halon 1211 Fire Extinguishing Systems*. Though the decomposition products generated when halons are exposed to flame or to temperatures in excess of 900°F have a relatively higher toxicity than the natural vapor, studies indicate that little, if any, risk is attached to the use of Halon 1301 or Halon 1211 when used in accordance with NFPA standards.

Application Methods

Halogenated extinguishing agents are discharged from total flooding and local application systems and from portable extinguishers (see Chapter 30).

Total Flooding Systems: These systems are designed to establish a specific extinguishing agent concentration in a given enclosure — in a computer room for example. It is important that the agent concentration be uniform throughout the enclosure. Total flooding can be achieved with modular or central storage systems.

In modular systems, a nozzle is connected to a halon container with little or no piping. Modules are placed throughout the space to be protected in numbers and locations that will result in the desired uniform agent concentration.

In central storage systems, cylinders containing halon are connected to a manifold, and the gas is delivered to fire nozzles through a piping network.

Local Application Systems: In these systems, nozzles are located so that, when the system is activated, the object to be protected is surrounded by a high concentration of agent.

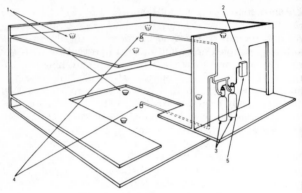

1. Automatic fire detectors installed both in room proper and in underfloor area.
2. Control panel connected between fire detectors and cylinder release valves.
3. Storage containers for room proper and underfloor area.
4. Discharge nozzles installed both in room proper and in underfloor area.
5. Control panel may also sound alarms, close doors, and shut off power to the area.

Figure 29-2. A total flooding halogenated extinguishing agent system installed in a room with a raised floor. (The Ansul Company)

Neither the quantity of agent nor the placement of nozzles is sufficient to achieve total flooding of the enclosure containing the object to be protected. Examples of equipment and facilities protected by local application are printing presses, dip tanks, and spray booths.

System Actuation

To limit the size and severity of a fire with which the system must deal, and thus minimize decomposition of the agent during extinguishment, automatic actuation with sensitive detection is recommended. Detectors must be sensitive enough to respond rapidly to fire in its early stages, but not so sensitive that they produce needless actuations.

Designers have used multiple detectors in a "cross zone" pattern to improve reliability. Actuation of either zone sounds local and remote alarms, but does not discharge the extinguishing agent. Actuation of both zones triggers discharge of the agent. The two zones may contain identical or different detectors.

In another arrangement, one zone of ionization detection provides early warning, and one zone of rate-compensated thermal detection causes the extinguishing agent to be released.

Maintenance

Seminannually, halon systems should be inspected visually for corrosion and damage and the storage containers checked for loss of agent. Containers are subjected to two tests, as neither alone is sufficient. First, pressure, corrected for temperature, should be measured to ensure no loss of pressurizing gas. Second, each container must be weighed to determine the loss of agent. In large systems, liquid level indicators may be provided, making weighing of the containers unnecessary.

At least annually, the system should receive a thorough visual examination, and the performance of detection, actuation, and alarm equipment should be tested with the agent release mechanism disconnected.

Annual and semiannual tests are usually conducted by the equipment manufacturer or installation contractor. If you do not actually witness the tests, you should at least be made aware that they were conducted and have access to the reports.

DRY CHEMICAL SYSTEMS

Dry chemical extinguishing agents are known as regular or ordinary dry chemicals and multipurpose dry chemicals. The former are used to combat fires involving flammable liquids (Class B) and electrical equipment (Class C); the latter are effective on ordinary combustibles (Class A) as well as on flammable liquids and electrical equipment.

Typical dry chemical agents use potassium bicarbonate, sodium bicarbonate, monoammonium phosphate, potassium chloride, or urea-potassium bicarbonate as a base material.

Dry chemical extinguishing agents should not be confused with dry powder agents, which were developed for use on combustible metals. Extinguishing agents used on combustible metal fires are discussed in Chapter 18.

Extinguishing Mechanism

When dry chemical agents are introduced directly into the fire area, extinguishment is almost instantaneous. Smothering, cooling, and radiation shielding contribute to the efficiency of dry chemical, but studies suggest that the principal mechanism of extinguishment is a chemical chain-breaking reaction in the flame.

Limitations

Dry chemical does not produce a lasting inert atmosphere above the surface of a flammable liquid; consequently, if ignition sources are present, flammable liquid vapors could be reignited.

Dry chemical should not be used where delicate electrical contacts, switches, and relays are present, as the insulating properties of the chemical may render the equipment inoperative. Some dry chemicals are corrosive and, for that reason, should be removed from undamaged surfaces soon after extinguishment.

Regular dry chemical will not extinguish fires that have penetrated below the surface, nor will dry chemical extinguish fires in materials that supply their own oxygen for combustion.

Hazards

Normally, dry chemical extinguishing agents are stable

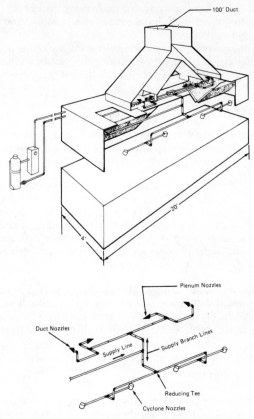

Figure 29-3. A typical single 30-pound cartridge-operated dry chemical system with fusible links for automatic operation, engineered for kitchen range, hood, duct, and fryer fire protection. (The Ansul Company)

materials. However, if various dry chemicals are mixed indiscriminantly, a dangerous reaction may occur. For example, if an acidic material is mixed with an alkaline material, carbon dioxide will be released and caking will occur. Portable extinguisher shells have been known to explode because of this phenomenon.

The ingredients of dry chemical agents are nontoxic, but they may cause temporary breathing difficulties and impair vision during and immediately after discharge.

Application Methods

Dry chemical extinguishing agents are discharged by an expellant gas from fixed systems, hand hose line systems, and portable extinguishers (see Chapter 30).

Fixed Systems: The components of fixed systems include the supply of dry chemical, an expellant gas, actuating devices, fixed piping and nozzles. Fixed systems may be designed for total flooding or local application. Total flooding discharges a fixed amount of agent into an enclosed space or enclosure around the hazard. Local application systems discharge the extinguishing agent directly into the fire.

Hand Hose Line Systems: These systems consist of a supply of dry chemical and expellant gas, and one or more hose lines to deliver the chemical to the fire. Hose stations are connected to the agent supply either directly or through intermediate piping.

System Actuation

Fixed systems are actuated by automatic mechanisms that employ sensing or detection devices in the hazard area and automatic mechanical or electrical releases that initiate the flow of dry chemical, actuate alarms, and shut down process equipment.

Maintenance

Dry chemical systems and their ancillary equipment should

be inspected at least annually — more frequently where the hazard or general conditions warrant.

Expellant gas should be checked semiannually — a pressure check if the gas is nitrogen, or a weight check if the gas is carbon dioxide. In stored pressure systems, the pressure gage is checked to see that the pressure is within the operating range.

Semiannual checks of the quantity of dry chemical should be made by looking at the agent level in separate dry chemical chambers or by weighing the stored pressure chamber. The dry chemical should be examined annually for evidence of caking in all but stored pressure systems.

During periodic inspections, check nozzles to see that they are properly aimed, free of obstructions, and in good condition. See that nozzles in kitchen range hood and duct systems are fitted with grease seals that are tight and easily blown clear. Examine actuating devices, such as fusible links, pneumatic heat detectors, or electric thermostats to see that they are not coated with residues or otherwise impaired.

The frequency of inspection of hand hose line systems will vary with location and climatic conditions. Equipment located in extremely hot or humid areas will require more frequent inspection, because heat can cause expellant gas pressure to increase and possibly cause leakage. When inspecting hand hose line systems, check the pressure of the expellant gas container or the unit itself if it is of the pressurized type. Inspect hoses and nozzles for obstructions and general condition.

INSPECTION OF SPECIAL EXTINGUISHING SYSTEMS

Experience is required for complete inspection of these systems. A few general points for the inspector are listed following:

(a) Inquire when the system was last inspected by the installer. The property owner should have a contract for inspection and testing at regular intervals.

(b) Determine if there has been any change in the hazard

protected since the system was designed and installed.

(c) Look at the equipment and devices for evidence that the various components of the system are suitable for their respective functions. The listings and markings of Underwriters Laboratories Inc., Underwriters Laboratories of Canada and Factory Mutual Engineering Corporation are evidence of the suitability of the special components of these systems. Carbon dioxide and dry chemical assemblies are usually listed as complete systems.

(d) Determine that discharge outlets are properly positioned and unplugged by dirt or residue.

(e) Determine that component equipment is located so as not to be needlessly subjected to fire exposure and that guards to prevent tampering are included where needed.

(f) Assure that piping and equipment are located and mounted so as to be mechanically secure and accessible for cleaning and maintenance.

(g) Determine the amount of the extinguishing agent provided for carbon dioxide, dry chemical, and halon systems. Check for fluidity of dry chemical periodically.

BIBLIOGRAPHY

McKinnon, G. P. ed., *Fire Protection Handbook*, 15th ed., National Fire Protection Association, Quincy, MA, 1981. Carbon dioxide, halogenated agent, and dry chemical agent extinguishing systems are treated in considerable detail in Chapters 1, 2, and 3 of Section 18.

NFPA Codes, Standards, and Recommended Practices. (See the latest *NFPA Codes and Standards Catalog* for the availability of current editions of the following documents.)

NFPA 12, *Standard on Carbon Dioxide Extinguishing Systems.* A standard on total flooding and hose line systems.

NFPA 12A, *Standard on Halon 1301 Fire Extinguishing Systems.* Installing, testing, inspection, and maintaining systems using Halon 1301.

NFPA 12B, *Standard on Halon 1211 Fire Extinguishing Systems.* A standard for the design and installation of systems using bromochlorodifluoromethane.

NFPA 17, *Standard for Dry Chemical Extinguishing Systems*. Total flooding, local application and hand hose dry chemical extinguishing systems.

NFPA 75, *Standard on Electronic Computer/Data Processing Equipment*. Standard requirements for installations needing fire protection or special building construction, rooms, areas, or operating environments.

Chapter 30

PORTABLE FIRE EXTINGUISHERS

Fires in their early stages can be easily extinguished with the application of the proper type and amount of extinguishing agent. Portable fire extinguishers are designed for this purpose, but their successful use depends upon several conditions.

• The fire must be discovered while it is still small enough to be extinguished by a portable unit.

• The extinguisher location must be obvious and accessible.

• The extinguisher must be of the proper type and capacity for the fire in progress, and it must be in operating condition.

• The person discovering the fire is trained and practiced in the use of the equipment.

CLASSIFICATION OF FIRES

Different extinguishing agents are required for different types of fuel. For convenience in discussing the suitability of various available extinguishing agents, NFPA 10, *Portable Fire Extinguishers*, classifies fires into four types.

Class A: Fires in ordinary combustible materials (such as wood, cloth, paper, rubber, and many plastics) which require the heat-absorbing (cooling) effects of water or water solutions, the coating effects of certain dry chemicals which retard combustion, or the interrupting of the combustion chain reaction by halogenated agents.

Class B: Fires in flammable or combustible liquids, flammable gases, greases, and similar materials, which must be put out by excluding air (oxygen), inhibiting the release of combustible vapors, or interrupting the combustion chain reaction.

Class C: Fires in live electrical equipment; safety to the operator requires the use of electrical nonconductive extinguishing agents (Note: when electrical equipment is de-energized, extinguishers for Class A or B fires may be used.)

Class D: Fires in certain combustible metals (such as magnesium, titanium, zirconium, sodium, potassium, etc.) which require a heating-absorbing extinguishing medium that does not react with the burning metals.

Some portable extinguishers will put out only one class of fire, and some are suitable for two or three, but none is suitable for all four. Most extinguishers are labeled so that users may quickly identify the class of fire for which they may be used.

Rating numerals are also used on the labels on extinguishers for Class A and Class B fires; the rating numeral gives the relative extinguishing effectiveness of the extinguisher. For example, an extinguisher rated for Class A fires has a rating number that precedes the letter "A"; this numeral gives the relative extinguishing capacity of the extinguisher as determined by standard reproducible test fires. An extinguisher with a 4-A rating can put out more fires involving ordinary combustibles than one with a 2-A rating, but not necessarily twice as much; the ratings are only relative.

Extinguishers rated for Class B fires also have numeral ratings, in this case based on the realtive quantity of burning flammable liquid in a flat pan that can be extinguished during a laboratory test. Again, the point is that an extinguisher rated 20-B can put out much more fire than one rated 5-B.

No rating numerals are used for extinguishers labeled for Class C fires. Since electrical equipment has either ordinary combustibles or flammable liquids, or both, as part of its construction, an extinguisher for Class C fires should be chosen according to the nature of the combustibles in the immmediate area.

Extinguishers for Class D fires contain different dry powders that are effective on fires in different kinds of combustible metals. One extinguisher for magnesium fires may not work on a sodium fire, or at least not with the same effec-

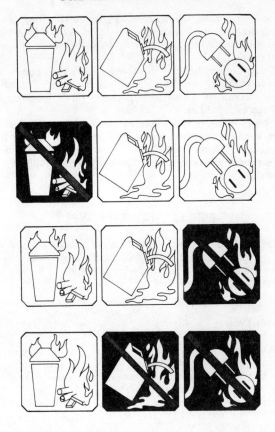

Figure 30-1. These pictographs are designed so that the proper use of an extinguisher can be determined at a glance. When an application is prohibited, the background is black and the slash is bright red. Otherwise, the background is light blue. Top row indicates an extinguisher for Class A:B:C fires; second row, Class B:C fires; third row, Class A:B fires; and fourth row, Class A fires.

tiveness. For that reason, general numerical ratings are not used; instead, each extinguisher for Class D fires has a nameplate detailing the type of metal the particular agent will extinguish.

Extinguishers that are effective on more than one class of fire have multiple "letter" and "numeral-letter" classifications and ratings. These are both shown on the labels applied to each extinguisher.

The most recently recommended marking system is one that combines pictographs of both uses and nonuses on a single label. Letter-shaped symbol markings are recommended for use until conversion to the newer pictographs is completed.

FIRE EXTINGUISHER TYPES

Currently listed fire extinguishers are classified into six major groups based on the extinguishing medium each contains. They are (1) water-type, (2) carbon dioxide, (3) halogenated agent, (4) dry chemical, (5) dry powder, and (6) foam.

Water Type

Water-based extinguishing agents include water, antifreeze, loaded stream, wetting agent, soda acid, and foam. All except AFFF foam are for use on Class A fires only. Soda-acid and ordinary foam extinguishers are no longer manufactured, but some may still be in use. The obsolete extinguishers are the type that must be inverted or inverted and bumped to initiate discharge. Water-type extinguishers being manufactured today are either the stored pressure or pump tank type.

Carbon Dioxide

These extinguishers are intended for use on Class B and Class C fires, but can be used on a Class A fire until a more

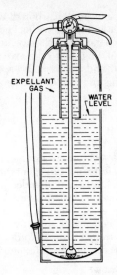

Figure 30-2. A stored-pressure water extinguisher.

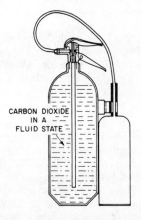

Figure 30-3. A carbon dioxide fire extinguisher.

suitable agent can be located. Carbon dioxide is stored in the extinguisher cylinder as a liquid with a vapor space at the top and is discharged as a gas. It is self-expelling. The discharged gas chills the surrounding air to the extent that moisture in the air is condensed and takes on the appearance of snow. Carbon dioxide extinguishers require no cold weather protection.

Halogenated Agent

Halon 1211 (bromochlorodifluoromethane) is similar to carbon dioxide in that it is a clean agent that leaves no residue to be cleaned up. Halon 1211 is intended primarily for use on Class B:C fires. However, extinguishers in sizes of 9 pounds capacity or greater are given a Class A rating. Although Halon 1211 is retained under pressure in a liquid state, a booster charge of nitrogen is usually added to ensure proper operation. Like carbon dioxide, Halon 1211 needs no cold weather protection.

Dry Chemical

There are two basic kinds of dry chemical agents used in portable extinguishers — ordinary dry chemicals for use on Class B:C fires and multipurpose dry chemicals for use on Class A:B:C: fires. Two discharge methods are used — stored pressure and expelling agent cartridge.

Dry Powder

Dry powder extinguishers are intended for use on Class D fires — combustible metals. Traditionally, dry powder has been applied to fires involving combustible metals with a scoop and shovel. There is a cartridge-operated portable dry powder extinguisher available, but it should not be used where there is danger of the agent stream spreading the burning material.

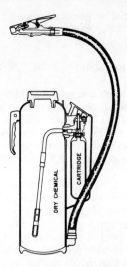

Figure 30-4. A cartridge-operated dry chemical fire extinguisher.

Aqueous Film-forming Foam

This agent is an aqueous film-forming surfactant in water which forms a mechanical foam when discharged through an aspirating nozzle. On Class A fires, the agent cools and penetrates to reduce temperatures to below ignition levels. On Class B fires, it acts as a barrier to exclude oxygen from the surface of the fuel. AFFF is available in 2½-gallon stored-pressure extinguishers.

DISTRIBUTION OF EXTINGUISHERS

When placing extinguishers, select location that will (1) provide uniform distribution, (2) provide easy access and be relatively free from temporary blockage, (3) be near normal paths of travel, (4) be near exits and entrances, (5) be free from the potential of physical damage, and (6) be readily available.

Mounting Extinguishers

Most extinguishers are mounted to walls or columns by securely fastened hangers so that they are adequately supported. Some extinguishers are mounted in cabinets or wall recesses. When this is the case the operating instructions should face outward and the extinguisher should be placed so that it can be easily removed. Cabinets should be kept clean and dry.

Where extinguishers may become dislodged, brackets specifically designed to cope with this problem are available. In areas where they are subject to physical damage (such as warehouse aisles), protection from impact is important. In large open areas (such as aircraft hangars), extinguishers may be mounted on movable pedestals or wheeled carts. In order to maintain some pattern of distribution and specify intended placement, locations should be marked on the floor.

The NFPA Fire Extinguisher Standard specifies floor clearance and mounting heights, based on extinguisher weight, as follows:

1. Extinguishers with a gross weight not exceeding 40 pounds should be installed so that the top of the extinguisher is not more than 5 feet above the floor.

2. Extinguishers with a gross weight greater than 40 pounds (except wheeled types) should be installed so that the top of the extinguisher is not more than 3½ feet above the floor.

3. In no case shall the clearance between the bottom of the extinguisher and the floor be less than 4 inches.

When extinguishers are mounted on industrial trucks, vehicles, boats, aircraft, trains, etc., special mounting brackets (available from the manufacturer) should be used. It is important that the extinguisher be located at a safe distance from the hazard so that it will not become involved in the fire.

Distribution for Class A Combustibles

Table 19-2B of the Fire Protection Handbook is a guide to determining the minimum number and rating of extinguishers for Class A fires needed in any particular area. Sometimes extinguishers with ratings higher than what the table indicates may be necessary because of process hazards, building configuration, etc., but in no case should the recommended maximum travel distance be exceeded.

The first step when calculating how many Class A extinguishers are needed is to determine whether an occupancy is light, ordinary, or extra hazard. (See NFPA 10.) Next, the extinguisher rating should be matched with occupancy hazard to determine the maximum area that an extinguisher can protect. Table 19-2B also specifies the maximum travel distance (actual walking distance) allowed; for Class A extinguishers it is 75 feet. For example, each 2½-gallon stored-pressure water extinguisher rated 2-A will protect an area of 3,000 square feet in an ordinary hazard occupancy, but only 2,000 square feet in an extra hazard occupancy.

The NFPA Extinguisher Standard also provides that up to one-half of the complement of extinguishers for Class A fires may be replaced by uniformly spaced small hose (1½-inch) stations. The hose stations and extinguishers, however, should be located so that the hose stations do not replace more than every other extinguisher.

Distribution for Class B Combustibles

In areas where liquids do not reach an appreciable depth, extinguishers should be provided according to Table 19-2C of the Fire Protection Handbook. The reason why the basic maximum travel distance to Class B extinguishers is 50 feet, as opposed to 75 feet for Class A extinguishers, is that flammable liquids fires reach their maximum intensity almost immediately, and thus the extinguisher must be nearer to hand. With lower-rated extinguishers, the travel distance drops to 30 feet.

When flammable liquids do reach an appreciable depth,

the extinguisher's rating number should be (except for foam types) at least twice the number of square feet of surface area of the largest tank in the area (assuming that other requirements are met). The travel distance specified by Table 19-2C should also be used to locate extinguishers for protection of spot hazards. Sometimes one extinguisher can even be installed to provide protection against several different hazards, provided that travel distances are not exceeded.

Distribution for Class C Fires

Extinguishers for Class C fires are required wherever there is live electrical equipment. This sort of extinguisher contains a nonconducting agent, usually carbon dioxide, dry chemical, or Halon 1211.

Once the power to live electrical equipment is cut off, the fire becomes a Class A, Class B, or Class A:B fire, depending on the nature of the burning electrical equipment and the burning material in the vicinity. Extinguishers for Class C fires should be selected according to (1) the size of the electrical equipment, (2) the configuration of the electrical equipment (particularly the enclosures of units, which influence agent distribution), and (3) the range of the extinguisher's stream. At large installations of electrical equipment where continuity of power is critical, fixed fire protection is desirable. But even when fixed fire protection is present, it is recommended that some Class C extinguishers be provided to handle incipient fires.

Class D Extinguisher Distribution

It is particularly important that the proper extinguishers be available for Class D fires. Because the properties of combustible metals differ, even an agent for Class D fires may be hazardous if used on the wrong metal. Agents should be carefully chosen according to the manufacturer's recommendations; the amount of agent needed is normally figured according to the surface area of the metal plus the shape and form of the metal, which could contribute to the severity of the fire and cause "bake-off" of the agent. For example, fires in magnesium filings are more difficult to put out than fires in magnesium scrap, and thus more agent is needed for magnesium filings. The maximum travel distance to all extinguishers for Class D fires is 75 feet.

INSPECTION OF EXTINGUISHERS

For each type of extinguisher, read carefully the manufacturer's instructions regarding periodic examination and maintenance. The following is a general checklist which may be helpful to inspectors.

1. Is the extinguisher clean and well cared for?

2. Has it been charged and hydrostatically tested within the prescribed periods and tagged to show the dates?

3. If a seal is provided (as in industrial and mercantile establishments and places of assembly where sealing may be required by the management or by law), is the seal intact?

4. Is the discharge orifice clear and unobstructed?

5. Is there indication that the cap, if any, may be cross-threaded on the collar or that threads are corroded?

6. Is the shell of the extinguisher corroded, damaged, or dented in such a way as to suggest possible weakness?

7. Are connections between the hose and the shell and nozzle secure?

8. If the extinguisher is a pump-operated type, does the pump shaft operate freely?

9. Is the location of the extinguisher readily accessible and plainly indicated so as to be visible at a distance?

10. If of a type subject to freezing, is the extinguisher protected?

11. Is hanger fastened solidly so that the extinguisher is well supported?

12. Is extinguisher located too close to the hazard which it is to protect, so that it could not be reached in case of fire?

SI UNITS

The following conversion factors are given as a convenience in converting to SI units the English units used in this chapter.

$$1 \text{ sq ft} = 0.0929 \text{ m}^2$$
$$1 \text{ in.} = 25.4 \text{ mm}$$
$$1 \text{ ft} = 0.305 \text{ m}$$
$$1 \text{ lb} = 0.454 \text{ kg}$$
$$1 \text{ gal} = 3.785 \text{ litres}$$

BIBLIOGRAPHY

McKinnon, G. P. ed., *Fire Protection Handbook*, 15th ed., National Fire Protection Association, Quincy, MA, 1981. Chapters 1 through 3 of Section 19 discuss the role, operation, distribution, inspection, and maintenance of portable fire extinguishers.

NFPA Codes, Standards, and Recommended Practices. (See the latest *NFPA Codes and Standards Catalog* for the availability of current editions of the following documents.)

NFPA 10, *Standard for Portable Fire Extinguishers*. Criteria for the selection, installation, inspection, maintenance, and testing of portable fire extinguishers.

INDEX

NO POSTAGE
NECESSARY
IF MAILED
IN THE
UNITED STATES

BUSINESS REPLY CARD

FIRST CLASS   PERMIT NO. 5347   QUINCY, MASS.

POSTAGE WILL BE PAID BY ADDRESSEE

NATIONAL FIRE PROTECTION ASSOCIATION

ATTN: PUBLICATIONS SALES DIVISION

BATTERYMARCH PARK

QUINCY, MA 02269

More NFPA resources...for the informed fire inspector.

Conducting Fire Inspections: A Guidebook for Field Use. The perfect companion book to your NFPA *Inspection Manual. Conducting Fire Inspections* gives you the facts you need to fine tune your inspection skills in various occupancies. Includes step-by-step coverage of 13 groups of occupancies as defined in the *Life Safety Code.* New loose-leaf format lets you record and summarize your observations and add extra notes as needed. Order your copy today. Approx. 220 pp. vinyl binder. (Order by No. SPP-75) $16.00 each

Life Safety Code® Handbook. The only handbook of its kind that explains and interprets the provisions of the 1981 *Life Safety Code.* Contains the complete text of the *Code* along with two-color commentary, useful cross references and detailed diagrams and illustrations throughout. Commentary contributed by members of the NFPA Committee on Safety to Life and by NFPA Code Specialists. 704 pp., 6½ x 9½. 1981. (Order by No. 101HB81) $19.50 each

ACT NOW!

Order these guides without delay. Mail this card, or use our easy direct dial order number — (617) 328-9230.

Please send me:

____ copy(s) of *Conducting Fire Inspections* 41 SPP-75
@ $16.00 each. $_____

____ copy(s) of the *Life Safety Code Handbook*
41-101HB81 @ $19.50 each. $_____

GRAND TOTAL $_____

☐ Payment Enclosed. ☐ Bill me.

NFPA Member No. _____
Payment with your order saves you handling charges. Prices subject to change without notice. Please allow 3 weeks for delivery.

Name _____

Organization _____

Address _____

City _____ State _____ Zip _____

National Fire Protection Association, Attn: Publications Sales Division
Batterymarch Park, Quincy, MA 02269